Max Plan

The Philosophy
of Physics

Translated by W. H. Johnston

With an Introduction by Michael J. Shaffer

MINKOWSKI
Institute Press

Max Karl Ernst Ludwig Planck
23 April 1858 – 4 October 1947

ISBN: 978-1-927763-62-9 (softcover)
ISBN: 978-1-927763-63-6 (ebook)

Minkowski Institute Press
Montreal, Quebec, Canada
http://minkowskiinstitute.org/mip/

For information on all Minkowski Institute Press publications
visit our website at http://minkowskiinstitute.org/mip/books/

PUBLISHER'S PREFACE

This is a new publication of Max Planck's book *The Philosophy of Physics.*[1]

Two main reasons make the book very valuable even in 21st century:

- it is written by the founding father of quantum physics

- the topics – Physics and World Philosophy, Causality in Nature, Scientific Ideas: Their Origin and Effects, Science and Faith – are as important now as they were at Planck's time.

As it had always been Planck's intention that his books and "essays should reach a wider circle of readers," this new publication should be of interest to physicists, philosophers of science, students and all interested in the foundations and philosophy of physics.

The book was typeset in LaTeX and noticed typos were corrected.

14 November 2019   Minkowski Institute Press

---

[1]Max Planck, *The Philosophy of Physics*, translated by W. H. Johnston (George Allen and Unwin, Ltd., London 1936).

ii

INTRODUCTION

Michael J. Shaffer
St. Cloud State University

Max Plank's *The Philosophy of Physics* was published in 1936, eighteen years after he received a Nobel Prize in Physics. He was awarded that distinguished honor, of course, for his fundamental contributions to physics in the form of the development of quantum mechanics. Given the title of his 1936 book and Planck's absolutely central role in the development of quantum mechanics one might think it would focus heavily, or even perhaps exclusively, on the technicalities of quantum mechanics and its direct physical implications more than on traditional philosophical matters. But, this is not the case. To be sure, in *The Philosophy of Physics* Planck does address the profound difference between classical mechanics and quantum mechanics. But, he also aims to elucidate a wide-ranging of matters having to do with truth, ethics, the character traits and biases of scientists, the nature of philosophical and scientific systems, the metaphysics of causality, the theory of measurement, the nature of good societies and even the nature of faith. This may seem to us today to be a case of a scientist over-stepping disciplinary boundaries in order to appear to be profound, but this is also not the case. Planck's discussion of more philosophical matters is central to the goal of this book.

In *The Philosophy of Physics* Planck was attempting to show how science, philosophy and society co-exist in an often-shaky balance that is in need of deeper understanding. In particular, Planck pays very close attention to important matters related to truth and bias as they manifest themselves in both science and in society more broadly. This is, of course, not surprising in light of Planck's having remained in Germany at the University of Berlin for the entirety of his career. Importantly, he was present in Germany to witness the rise of Nazism in the 1930s and the subsequent domination of German culture by a populist fascism that lasted until the end of World War II. To be clear, Planck was a staunch critic of Nazism who publicly stood in opposition to this anti-intellectual atrocity and this is evinced in what he has to say about the relationship between physics and philosophy. We shall return to this matter shortly, but first let us place Planck's book in its historical context more fully.

Planck's philosophical views importantly show the influence of the Berlin Circle headed by Hans Reichenbach and they reflect the move away from the prevailing Kantianism that antedated the rise of positivism in the early 20th century in Europe. His approach to philosophy and science is not a surprising one given that Planck himself directed Moritz Schlick's doctorate in Berlin and given that Planck, along with Einstein, also helped to secure a position for Reichenbach at the University of Berlin. Schlick founded the logical positivism that in turn affected Reichenbach's views and they could not avoided being influenced by Planck's own thinking given Planck's role in their academic lives.[2] So, there is little doubt that Planck's thinking was closely related to the work of the members of the Berlin Circle, especially Reichenbach's views. The Berlin Circle was greatly influential in the intellectual scene in Berlin at the time when Planck wrote *The Philosophy of Physics* and the view of science and philosophy

---

[2]Of course, Planck explicitly criticized aspects of positivism in Planck 1931 and later in Planck 1933. Schlick responded to Planck's 1931 criticisms in Schlick 1932.

that he espouses therein is very similar to the Reichenbach's views presented in his *The Theory of Relativity and A Priori Knowledge*, originally published in 1920. But, it also extends that view to a broader application.

In *The Philosophy of Physics* Planck sets the stage by noting that science and general philosophy cannot be entirely disentangled. This is simply because general philosophy concerns itself with everything and so must encompass physics and the rest of the sciences. Any reasonable philosophical system however must not conflict with what we have learned about nature via science, for otherwise it would go against our best knowledge of the nature of reality. So, the two domains are inter-dependent as Planck sees it. On this basis, Planck importantly sees that physics and the other sciences cannot be practiced in isolation from general philosophy. In other words, all physical theories presuppose some philosophical framework or principles of classification, but, *pace* Kant, there is no single a priori true framework or principle of classification that must be assumed as a matter of necessity. Such principles stand above the scientific theory and shape how we see and investigate the world, but there are many such frameworks that could be adopted. Moreover, Planck holds that the adoption of any philosophical framework requires making a value judgment concerning the appropriateness of that framework for the guidance of scientific research and as a set of presuppositions about both methodology and reality. This includes adopting familiar methodological values like respect for truth and commitments to principles like that of causality and the basic concept of a physical object. But, this also includes adopting more broad values like freedom of thought and inclusiveness. It is in virtue of this fact that Planck sees that every physical theory presupposes some philosophical theory and that values infect all science. It is also in virtue of this fact that Planck sees that biases in the form of framework assumptions can take root. They can both distort our views vi

of reality and be used to advance immoral aims.

Given this understanding of science, Planck asserts that many bitter scientific controversies are actually disputes about the selection of principles of classification or frameworks rather than about purely empirical matters. This is especially important because Planck believes that judgments concerning which philosophical framework to adopt are matters of convention guided by purely pragmatic implications. They are then effectively subjective biases that are not subject to empirical resolution. Different scientists or scientific communities can approach empirical inquiry differently in terms of different assumed frameworks grounded in value judgments even if they are not aware of this. Thus it is of the utmost importance *both* that scientists concern themselves with the search for truth and that scientists concern themselves with the search for correct values. There simply is no science practiced independently of philosophy, and specifically independently of both conventional principles of classification and values. In other words, scientists should not pretend that empirical science is value free and scientists should not pretend that science does not require philosophical support.

Planck's way of looking at science and its philosophical presuppositions then suggests that there can be importantly different kinds of conflicts between belief systems involving empirical theories. There can be disputes about the non-empirical philosophical frameworks associated with empirical theories and there can be disputes about empirical theories framed in terms of the same non-empirical philosophical framework. Importantly, as Planck sees it, the first kind of dispute can be resolved only by appeal to the pragmatic implications associated with the conflicting frameworks, whereas conflicts of the second sort can be resolved by appeal to the empirical basis of science (i.e. measurements). Where we have conflicts of the first sort we can then only look at the conventional frameworks adopted and then consider the consequences they entail and how we pragmatically value them. So, as we have already noted, this view involves the rejection of Kant's idea of the fixed and a priori warranted categories of though and the forms of sense and replaces it with the idea that all science is conducted in terms of some contingently adopted philosophical framework or other. But, the selection of any such framework is a non-empirical matter (i.e. a convention). So, Planck's view is very nearly identical to Reichenbach's view of what has come to be called the relativized a priori and its supposed role in the conduct of science. Michael Friedman (2001) has recently revived this sort of view. He understands Reichenbach's view of science from *The Theory of Relativity and A Priori Knowledge* usefully as follows:

> Reichenbach distinguishes two meanings of the Kantian a priori: necessary and unrevisable, fixed for all time, on the one hand, and "constitutive of the concept of the object of [scientific] knowledge," on the other. Reichenbach argues, on this basis, that the great lesson of the theory of relativity is that the former meaning must be dropped while the latter must be retained. Relativity theory involves a priori constitutive principles as necessary presuppositions of its properly empirical claims, just as much as did Newtonian physics, but these principles have essentially changed in the transition from the latter theory to the former... What we end up with, in this tradition, is thus a relativized and dynamic concept of a priori mathematical-physical principles, which change and develop along with the development of the mathematical and physical sciences themselves, but which nevertheless retain the characteristically Kantian constitutive function of making the empirical nat-

ural knowledge thereby structured and framed by
such principles first possible (2001, 30-31).

So, Reichenbach also held that all scientific theories are
presented in conjunction with a philosophical framework that
makes the empirical application of that theory possible and
that such philosophical frameworks are not selected on the
basis of empirical considerations.[3] In virtue of this methodological point, Planck demonstrates that at least some of the
opposition between classical physics and quantum mechanics
appears to be of this sort and has to do with the assumption of
the principle of causality. This means that the opposition between these theories is, at least in part, non-empirical and re-
solving such a problem unavoidably involves attending to the
philosophy behind the science. Planck then uses this model
of the methodology of philosophically framed physics as a
model for the conduct of all science and he warns us against
attempts to ignore the infection of value-based assumptions in
our thinking in all domains. As he sees it, we do so at our potential peril and we do so in such a way that we misrepresent
the methodology of the sciences. *The Philosophy of Physics*
is then an important and cautionary book and we would do
well to learn its lessons. Science needs philosophy and science
should be based on good values that neither pervert the truth
nor diminish our fellow men. As Planck explicitly warns us,

> Justice is inseparable from truthfulness: justice,
> after all, simply means the consistent application
> in practice of the ethical judgments which we pass
> on opinions and actions. The laws of nature re-
> main fixed and unchanged whether applied to great
> or to small phenomena, and similarly the commu-
> nal life of men requires equal right for all, for great
> and small, for rich and poor. All is not well with

---

[3]See Shaffer 2011 for critical discussion of this view.

> the State if doubts arise about the certainty of the
> law, if rank and family are respected in the courts,
> if defenseless persons feel that they are no longer
> protected from the rapacity of powerful neighbors,
> and if the law is openly wrenched on the grounds
> of so-called expediency (p. 19).

This is surely a sentiment that we would do well to take to heart.

References

Friedman, M. (2001). *Dynamics of Reason.* Stanford: CSLI Publishing.

Planck, M. (1931). *Positivismus und reale Aussenwelt.* Leipzig: Akademische Verlagsgesellschaft.

Planck, M. (1933). *Where is Science Going?* J. Murphy (trans.). London: George Allen and Unwin.

Schlick, M. (1932). "Positivismus und Realismus," *Erkenntnis* 3: 1–31.

Shaffer, M. (2011). "The Constitutive A Priori and Epistemic Justification," in *What Place for the A Priori?* M. Veber and M. J. Shaffer (eds.). Chicago: Open Court.

x

CONTENTS

1 PHYSICS AND WORLD PHILOSOPHY

The subject of this chapter is the connection between physics and the endeavor to attain a general philosophy of the world; and it may well he asked wherein this connection consists. Physics, it may be urged, is solely concerned with the objects and events of inanimate nature, while a general philosophy, if it is to be at all satisfactory, must embrace the whole of physical and intellectual life and must deal with questions of the soul, including the highest problems of ethics.

At first sight this objection may seem convincing. Yet it will not bear closer investigation. In the first place inanimate nature is, after all, part of the world, so that any philosophy of the world claiming to be truly comprehensive must take notice of the laws of inanimate nature; and in the long run such a philosophy becomes untenable if it conflicts with inanimate nature. I need not here refer to the considerable number of religious dogmas to which physical science has dealt a fatal blow.

The influence of physics upon a general world philosophy is not, however, confined to such a negative or merely destructive activity; its contribution in a positive sense is of much greater importance. This is true with regard both to form and to content. It is common knowledge that the methods of physical science have proved so fruitful largely on account of their exactness and have on this account provided a model for not strictly scientific studies; while in regard to content it should be said that every science has its roots in life and that similarly physics can never be completely separated from its student; every student, after all, is a personality equipped with a set of intellectual and ethical properties. Hence the general philosophy of the student will always have some influence on his scientific work, while conversely the results of his studies cannot but exert some influence on his general philosophy. It will be the chief purpose of the present chapter to demonstrate this in detail with respect to physics.

I propose to begin with a general consideration. Any scientific treatment of a given material demands the introduction of a certain order into the material dealt with: the introduction of order and of comparison is essential if the available and steadily increasing matter is to be grasped; and the obtaining of such a grasp is essential if the problems are to be formulated and pursued. Order, however, demands classification; and to this extent any given science is faced by the problem of classifying the available material according to some principle. The question then arises, what is to be this principle? Its discovery is not only the first but, as ample experience proves, frequently the decisive step in the development of any given science.

It is important at this point to state that there is no one definite principle available *a priori* and enabling a classification suitable for every purpose to be made. This applies equally to every science. Hence it is impossible in this connection to assert that any science possesses a structure evolving from its own nature inevitably and apart from any arbitrary presupposition. It is important that this fact should be clearly grasped; it is of a fundamental significance because it demonstrates that it is essential, if there is to be any scientific knowledge, to determine the principle in accordance with which its studies are to be pursued. This determination cannot be made merely in accordance with practical considerations; questions of value also play their part.

Let us take a simple example from the most mature and exact of all sciences, mathematics. Mathematics deals with the magnitude of numbers. In order to obtain a survey of all numbers the obvious method would be to classify them by magnitude; in which case any two numbers are close to each other in proportion as the difference between them is small. Let us take two numbers which are practically equal in magnitude, one of them being the square root of 2 and the other 1.41421356237. The former figure is a few billionths greater than the latter and in every numerical calculation in physics or in astronomy the two numbers can be treated as completely identical. So soon, however, as numbers are classified in accordance with their origin, and not in accordance with their magnitude, a fundamental difference between the two numbers arises. The decimal fraction is a rational number and can be expressed by the ratio between two integers, while the square root is irrational and cannot be so expressed. If now it is asked whether these two numbers are closely related to each other or not, then any dispute on this question formulated in this manner would have no more meaning than a dispute between two persons facing each other and debating which side was right and which left. I have taken this simple example because I am convinced that many scientific controversies, and among them many which aroused a maximum of bitterness, have ultimately been due to the fact that the two opponents were, without clearly stating it, employing different principles of classification in the arrangement of their arguments. Every kind of classification is inevitably vitiated by a certain element of caprice and hence of onesidedness. The selection of the principle of classification is even more important in the natural sciences. As an example one might take botany. Some kind of nomenclature is essential and hence all plants must be divided according to species, genera, families, etc.

But according as different principles of classification were selected, so different systems evolved. In the history of botany there have sometimes been sharp controversies between these systems, none of which can claim infallibility since each is affected by subjective bias. The natural system of plants now in general use, although superior to the earlier artificial systems, is not definitive nor clearly determined in every detail, but is subject to certain fluctuations corresponding to the different attitudes taken by leading investigators to the question of the most expedient principle of classification.

The necessity of introducing some classification and the caprice attaching to it is most striking and significant, however, in the non-scientific studies and especially in history. Whether history is classified vertically or horizontally, whether it is arranged according to political, ethnographic, linguistic, social, or economic principles, the necessity continually arises of making distinctions which are seen on close consideration to be fluid and inadequate for the simple reason that any kind of classification inevitably separates cognate subjects and sunders closely allied matters. Thus every science contains an element of caprice and hence of transitoriness in its very structure, a defect which cannot be eradicated because it is rooted in the nature of the case.

In turning to physics we are now faced by the task of classifying under various groups the events which we study. This much is a preliminary demand. Now all physical experiences are based upon our sense-perceptions, and accordingly the first and obvious system of classification was in accordance with our senses. Physics was divided into mechanics, acoustics, optics, and heat. These were treated as distinct subjects. In course of time, however, it was seen that there was a close connection between these various subjects, and that it was much easier to establish exact physical laws if the senses are ignored and attention is concentrated on the events outside the senses – if, for example, the sound waves emanating from a sounding body are dealt with apart from the ear, and the rays of light emanating from a glowing body apart from the eye. This leads to a different classification of physics, certain parts of which are re-arranged, while the organs of sense recede into the background. According to this principle the heat rays emanating from a hot stove ceased to be the province of heat and were assigned to optics, where they were dealt with as though entirely similar to light waves. Admittedly such a re-arrangement, neglecting as it does the perceptions of the senses, contains an element of bias and arbitrariness. Goethe, who always insisted on the primacy of the senses, would have been horrified by such an arrangement; for Goethe always concentrated on the event in its totality, insisted on the superiority of the immediate sensation and hence would never have agreed to a distinction between the organ of sight and the source of light.

> If the eye were not of the nature of the sun How could we see the light?

Yet it may be presumed that, had he lived a century later, Goethe would not have objected to the soothing light of an electric bulb on his desk, although its invention was made possible only by the particular physical theory which he had so vigorously opposed. Neither Goethe nor his great adversary Newton could have suspected while alive that this successful theory when consistently developed was doomed to give way to the opposite onesidedness. I do not wish to anticipate, however, and now revert to a description of the further development of physics. Once the specific perceptions of the senses as fundamental concepts of physics had been eliminated from that science, it was a logical step to substitute suitable measuring instruments for the organs of sense. The eye gave way to the photographic film, the ear to the vibrating membrane, and the skin to the thermometer. The introduction of self-registering apparatus further eliminated subjective sources of error. The essential characteristic of this development, however, did not consist in the introduction of new measuring instruments of steadily growing sensitiveness and exactitude: the essential point was that the assumption that measurement gave immediate information about the nature of a physical event – whence it followed that the events were independent of the instruments used for measuring them – now became the foundation of the theory of physics. On this assumption a distinction must be made, whenever a physical measurement takes place, between the objective and actual event, which takes place completely independently, and the process of measuring, which is occasioned by the event and renders it perceptible. Physics deals with the actual events, and its object is to discover the laws which these events obey.

This method of interrogating nature has been justified in the past by the wealth of results obtained by classical physics; for classical physics followed the methods indicated by this view and the results applied in practical life to applied science and to kindred pursuits are familiar and visible to all. A detailed description is hence unnecessary.

Encouraged by this success physicists proceeded on the road which they had entered. They continued to apply the principle of *divide et impera*. After the actual events had been separated from the measuring instruments bodies were divided up into molecules, molecules into atoms, and atoms into protons and electrons. Simultaneously space and time were divided into infinitely small intervals. Everywhere rigorous laws were sought and found; as the process of sub-division went on, so the laws assumed simpler forms and there seemed to be no reason for not assuming that it might prove possible to reduce the laws of the physical macrocosm to the same spatial-temporal differential equations which are valid for the microcosm. These equations would then give for any given initial state of nature the recurring changes and hence by integration the states for all future time, a view of the physical events of the world as comprehensive as it was satisfactory by reason of its harmony.

The surprise was all the more striking and unpleasant when, at the beginning of the present century, the increasing delicacy and number of available methods of measurement showed, first in the field of heat radiation, later in that of light rays, and finally in that of electromechanics, that the classical theory as described above is faced by an insurmountable barrier. It may be best to give an example. In order to calculate the movement of an electron, classical physics must assume that its state is known, and this state embraces its position and its velocity. Now it was found that every method permitting of an exact measurement of the electron's position prohibits an exact measurement of its velocity: and it was further found that the inaccuracy of the latter measurement varies inversely with the accuracy of the former, and vice versa, the phenomenon being governed by a law which is accurately defined by the magnitude of Planck's quantum. If the position of the electron is known exactly its velocity is not known at all, and vice versa.

Clearly in these circumstances the differential equations of classical physics lose their fundamental importance; and for the time being the task of discovering in all their details the laws underlying the real physical processes must be regarded as insoluble. But of course it would be incorrect to infer that no such laws exist: the failure to discover a law will, on the contrary, have to be attributed to an inadequate formulation of the problem and a consequently incorrect posing of the question. The question now is wherein the mistake consists and how it can be removed.

It should be stressed first that it would be incorrect to speak of a breakdown of theoretical physics in the sense that everything achieved hitherto must be regarded as incorrect and must hence be rejected. The successes attained by classical physics are far too important to permit such drastic

action. It is not the case that a new structure has to be
erected, but that an old theory must be extended and elabo-
rated, this being true especially with regard to micro-physics;
in the field of macro-physics, which deals with relatively large
bodies and spaces of time, the classical theory will always retain its importance. Clearly, then, the mistake does not lie
in the fundamentals of the theory but in the fact that among
the assumptions used for building it up there must be one to
which the failure is due, the elimination of which would allow
the theory to be further extended.

Let us consider the facts of reality. Theoretical physics
is based on the assumption that there exist real events not
depending upon our senses. This assumption must in all circumstances be maintained; and even physicists of positivist
leanings make use of it. Even if this school maintains that the
priority of the sense data is the sole foundation of physics, it
is yet compelled, in order to escape an irrational solipsism,
to assume that there are such things as individual deceptions
of the senses and hallucinations; and these can be eliminated
only on the assumption that physical observations can be re-
produced at will. This, however, implies what is not evident
a priori, namely, that the functional relations between sense
data contain certain elements not depending upon the observer's personality nor upon the time and place of observation. It is precisely these elements which we describe as the
real part of the physical event and of which we attempt to
discover the laws.

We saw above that classical physics, besides assuming the
existence of real events, has always further assumed the possibility of obtaining a complete grasp of the laws governing the
real events, the method of obtaining this grasp being a pro-
gressive, spatial and temporal sub-division in the direction of
the infinitely small. More closely considered this assumption
must be largely modified, since it leads, e.g., to the conclusion that the laws governing a real event can be completely

understood if it is separated from the event by which it is measured. Now evidently the process of measuring can inform us about the real event only if there is some kind of causal connection between the two, and if there is such a connection, then the process of measuring will, in some degree, influence and disturb the event, with the consequence that the result of the measurement is falsified. This falsification and the consequent error will be great in proportion as the causal nexus between the real objective and the measuring instrument is close and delicate; it will be possible to reduce it by relaxing the causal nexus or, to express it differently, by increasing the causal distance between the object and the measuring instrument. It is never possible to eliminate the interference altogether, since, if the causal distance is assumed to be infinitely great, i.e., if we completely sever the object from the measuring instrument, we learn nothing at all about the real event. Now the measuring of single atoms and electrons requires extremely delicate and sensitive methods and hence implies a close causal nexus; the exact determination of the position of an electron therefore implies a relatively powerful interference with its motion; and conversely the exact measurement of the velocity of an electron requires a relatively lengthy time. In the first case there is interference with the electron's velocity; in the second, its position in space becomes indefinite. This is the causal explanation of the inaccuracy described above.

Convincing as these considerations may appear, they do not reach the core of the problem. The fact that a physical event is interfered with by the measuring instrument is familiar in classical physics; and at first it is not apparent why increasing improvements in methods of measuring should not permit us ultimately to calculate in advance the amount of the interference when dealing with electrons. If, therefore, we wish to understand the failure of classical physics in the microcosm, we must carry our investigations somewhat deeper.

The study of this question was carried forward considerably by the establishment of quantum mechanics or wave mechanics, from whose equations observable atomic processes can be calculated in advance. If the rules are observed the results of such calculation agree exactly with experience. It is true that, unlike classical mechanics, quantum mechanics does not give the position of an individual electron at any given time; what it does is to state the probability that an electron will be at a given place at a given time; or alternatively, given a multitude of electrons, it states the number which in any given time will be at a given place.

This is a law of a purely statistical character. The fact that it has been confirmed by all measurements hitherto made, and the further fact that there is such a thing as the uncertainty relation, has induced certain physicists to conclude that statistical laws are the only valid foundations of every physical law, more particularly in the field of atomic physics; and to declare that any question about the causality of individual events is, physically, meaningless.

We here reach a point whose discussion is of particular importance, since it leads us to a fundamental question: what is the task and what are the achievements of physics? If we hold that the object of physics is to discover the laws governing the relation between the real events of nature, then causality becomes a part of physics, and its deliberate elimination must give rise to certain misgivings.

It should first be observed that the validity of statistical laws is entirely compatible with a strict causality. Classical physics contains numerous examples. Thus, we may explain the pressure of a gas on the wall of the containing vessel as due to the irregular impingement of numerous gas molecules flying about in all directions; but this explanation is compatible with the admission that the impingement of anyone molecule upon another or upon the wall is governed by law and hence is completely determined causally. It may be objected that a strict causality can be regarded as definitely proved only if we are in a position to predict the entire course of the event; and it might be added that nobody can check the movement of any single molecule. To this we might reply that a rigorously exact prediction is never possible of any natural event, so that the validity of the law of causality can never be demonstrated by an immediate and exact experiment, since every measurement, however exact, inevitably involves certain errors of observation. Yet in spite of this the result of the measurement as well as individual errors of observation are attributed to definite causes. When we watch the waves breaking on the sea shore, we have every right to feel convinced that the movement of every bubble is due to strict causal law, although we could never hope to follow its rise and fall, still less to calculate it in advance.

It is at this point that the uncertainty relation is brought forward. While classical physics was fashionable, it might be hoped that the inevitable errors of observation could be reduced beneath any given limit by an appropriate increase in the accuracy of measurements. This hope was destroyed by the discovery of Planck's quantum, since the latter implies a fixed objective limitation of the exactitude which can be reached, within which limit there is no causality but only doubt and contingency.

We have already prepared a reply to this objection. The reason why the measurements of atomic physics are inexact need not necessarily be looked for in any failure of causality; it may equally well consist in the formulation of faulty concepts and hence of inappropriate questions.

It is precisely the reciprocal influence between the measurement and the real event which enabled us to understand the uncertainty relation at least to a certain degree. According to this view we can no more follow the movement of the individual electron than we can see a colored picture whose dimensions are smaller than the wave length of its color.

It is true that we must reject as meaningless the hope that

it might eventually prove possible indefinitely to reduce the inaccuracy of physical measurements by improving the instrument. Yet the existence of an objective limit like Planck's quantum is a sure indication that a certain novel law is at work which has certainly nothing to do with statistics. Like Planck's quantum every other elementary constant, e.g., the charge or mass of an electron, is a definite real magnitude; and it seems wholly absurd to attribute a certain fundamental inexactitude to these universal constants, as those who deny causality would have to do if they wish to remain consistent.

The fact that there is a limit to the accuracy of the measurements in atomic physics becomes further intelligible if we consider that the instruments themselves consist of atoms and that the accuracy of any measuring instrument is limited by its own sensitiveness. A weigh-bridge cannot weigh to the nearest milligramme.

Now what can we do if the best that we have is a weigh-bridge and there is no hope of obtaining anything more accurate? Would it not be better to give up hope of obtaining exact weights and to declare the pursuit of the milligramme to be meaningless, rather than to pursue a task which cannot be solved by direct measurement? This argument underestimates the importance of theory: for theory takes us beyond direct measurement in a way which cannot be foretold *a priori*, and it does so by means of the so-called intellectual experiments which render us largely independent of the defects of the actual instruments.

It is wholly absurd to maintain that an intellectual experiment is important only in proportion as it can be checked by measurement; for if this were so, there could be no exact geometrical proof. A line drawn on paper is not really a line but a more or less narrow strip, and a point a larger or smaller spot. Yet nobody doubts that geometrical constructions yield a rigorous proof.

The intellectual experiment carries the mind of the investigator beyond the world and beyond actual measuring instruments and enables him to form hypotheses and to formulate questions which, when checked by actual experiment, enable him to perceive new laws even when these do not admit of direct measurement. An intellectual experiment is not tied down to any limits of accuracy, for thoughts are more subtle than atoms or electrons, nor is there any danger that the event which is measured can be influenced by the measuring instrument. An intellectual experiment requires one condition only for its success, and this is the admission of the validity of any non-self-contradictory law governing the relations between the events under observation. We cannot hope to find what is assumed not to be existent.

Admittedly an intellectual experiment is an abstraction; an abstraction, however, as essential to the experimenter and to the theorist as the abstract assumption that there is a real external world. Whenever we observe an event taking place in nature we must assume that something is happening independently of the observer, and conversely we must endeavor to eliminate as far as possible the defects of our senses and of our methods of measurement in order to grasp the details of the event with greater perfection. There is a kind of opposition between these two abstractions: while the real external world is the object, the ideal spirit which contemplates it is the subject. Neither can be logically demonstrated and hence no *reductio ad absurdum* is possible if their existence is denied. The history of physics bears witness, however, that they have played a decisive part throughout its development. The choicest and most original minds, men like Kepler, Newton, Leibniz, and Faraday, were inspired by the belief in the reality of the external world and in the rule of a higher reason in and beyond it.

It should never be forgotten that the most vital ideas in physics have this two-fold origin. In the first instance the form which these ideas take is due to the peculiar imagination of the individual scientist: in course of time, however, they assume a more definite and independent form. It is true that there have always been in physics a number of erroneous ideas on which a quantity of labor was wasted: yet on the other hand many problems which were at first rejected as meaningless by keen critics were eventually seen to possess the highest significance. Fifty years ago positivist physicists considered it meaningless to ask after the determination of the weight of a single atom – an illusory problem not admitting scientific treatment. Today the weight of an atom can be stated to within its ten-thousandth part, although our most delicate scales are no more fit to weigh it than a weigh-bridge is to determine milligrammes. One should therefore beware of declaring meaningless a problem whose solution is not immediately apparent; there is no criterion for deciding *a priori* whether any given problem in physics has a meaning or not, a point frequently overlooked by the positivists. The only means of judging a problem correctly consists in examining the conclusions to which it leads. Now the assumption that there are rigid laws applicable to physics is of such fundamental importance that we should hesitate before we declare the question whether such laws are applicable to atomic physics to be a meaningless one. Our first endeavor, on the contrary, should be to trace out the problem of the applicability of laws in this field.

Our first step should be to ask why classical physics fails in the question of causality when the interference arising from the measuring instrument and the inadequate accuracy of the latter are both insufficient to explain this failure. Plainly we are forced to adopt the obvious but radical assumption that the elementary concepts of classical physics cease to be applicable in atomic physics.

Classical physics is based on the assumption that its laws are most clearly revealed in the infinitely small; for it assumes that the course of a physical event anywhere in the universe is completely determined by the state prevailing at this place and its immediate vicinity. Hence such physical magnitudes relating to the state of the physical event as position, velocity, intensity of the electric and magnetic field, etc., are of a purely local character, and the laws governing their relation can be completely expressed by spatial-temporal differential equations between these magnitudes. Clearly, however, this will not suffice for atomic physics, so that the above concepts must be made more complete or more universal. In which direction, however, is this to be done? Some indication may perhaps be found in the recognition, which is daily spreading wider, that the spatial-temporal differential equations do not suffice to exhaust the content of the events within a physical system and that the liminal conditions must also be taken into consideration. This applies even to wave mechanics. Now the field of the liminal conditions is always finite and its immediate interference in the causal nexus is a new manner of looking at causality and one hitherto foreign to classical physics.

The future will show whether progress is possible in this direction and how far it will lead. But whatever results it may ultimately reveal, it is certain that it will never enable us to grasp the real world in its totality any more than human intelligence will ever rise into the sphere of ideal spirit: these will always remain abstractions which by their very definition lie outside actuality. Nothing, however, forbids us to believe that we can progress steadily and without interruption to this unattainable goal; and it is precisely the task of science with its continual self-correction and self-improvement to work in this direction without cease once it has been recognized that it is a hopeful direction. This progress will be a real one and not an aimless zig-zag, as is proved by the fact that each new stage reached enables us to survey all the previous stages, while those which remain to be covered are still obscure; just as a climber trying to reach higher altitudes looks down upon the distance he has covered in order to gain knowledge for the further ascent. A scientist is happy, not in resting on his attainments but in the steady acquisition of fresh knowledge.

I have so far confined myself to physics; but it may be felt that what has been said has a wider application. Natural science and the intellectual sciences cannot be rigorously separated. They form a single inter-connected system, and if they are touched at any part the effects are felt through all the ramifications of the whole, the totality of which is forthwith set in motion. It would be absurd to assume that a fixed and certain law is predominant in physics unless the same were true also in biology and psychology. We may perhaps here deal with free will. Our consciousness, which after all is the most immediate source of cognition, assures us that free will is supreme. Yet we are forced to asked whether human will is causally determined or not. Put in this way the question, as I have frequently tried to show, is a good example of the kind of problem which I have described as illusory, by which I mean that, taken literally, it has no exact meaning. In the present instance the apparent difficulty is due to an incomplete formulation of the question. The actual facts may be briefly stated as follows. From the standpoint of an ideal and all-comprehensive spirit, , like every material and spiritual event, is completely determined causally. Looked at subjectively, however, the will, in so far as it looks to the future, is not causally determined, because any cognition of the subject's will itself acts causally upon the will, so that any definitive cognition of a fixed causal nexus is out of the question. In other words, we might say that looked at from outside (objectively) the will is causally determined, and that looked at from inside (subjectively) it is free. There is here no contradiction, any more than there was in the previous debate about the right- and left-hand side, and those who fail to agree to this overlook or forget the fact that the subject's will is never completely subordinate to its cognition and indeed always has the last word.

In principle, therefore, we are compelled to give up the attempt to determine in advance the motives guiding our actions on purely causal lines, i.e., by means of purely scientific cognition; in other words, there is no science and no intellect capable of answering the most important of all the questions facing us in our personal life, the question, that is, how we are to act.

It might thus be inferred that science ceases to play a part as soon as ethical problems arise. Yet such an inference would be wrong. We saw above that in dealing with the structure of any science, and in discussing its most suitable arrangement, a reciprocal inter-connection between epistimological judgments and judgments of value was found to arise, and that no science can be wholly disentangled from the personality of the scientist. Modern physics has given us a clear indication pointing in the same direction. It has taught us that the nature of any system cannot be discovered by dividing it into its component parts and studying each part by itself, since such a method often implies the loss of important properties of the system. We must keep our attention fixed on the whole and on the inter-connection between the parts.

The same is true of our intellectual life. It is impossible to make a clear cut between science, religion, and art. The whole is never equal simply to the sum of its various parts. And this is true also of mankind. It would be folly to attempt to obtain an understanding of mankind by studying a number of men however great; for each individual belongs to some community, to a family, a clan, or a nation – a community of which he must form a part, to which he must subordinate himself, and from which he cannot sever himself with impunity. For this reason every science, like every art and every religion, has grown up on a national foundation. It was the misfortune of the German people that this was forgotten for so many years.

It may be said that there is nothing new in this, and that it can be acknowledged without the aid of physics. This is true;

and all that I wish to show is that the position of physics, far from being unique, leads us to the same results and the same views as every other science, however different may be the point from which it starts. The real strength of its position is, in fact, seen if our argument is further developed; for it is only then that its tendency can be most clearly seen, which is to disregard its immediate origin and to expand in every direction like a healthily growing tree which tends to grow into the air and to stretch its branches in every direction, though at the same time it remains firmly rooted in the soil. If science is unable or unwilling to extend beyond the limits of the nation it is unworthy of the name of science; and in this connection physics enjoys an advantage over other branches of science. Nobody will dispute that the laws of nature are the same in every country; so that physics is not compelled to establish its international validity, unlike history where it has actually been asked whether an objective history can be an ideal to be aimed at. Ethics also is supranational, otherwise ethical relations could not exist between the members of different nations. Here again physics takes up a strong position. Scientifically it is based on the principle that it must contain no contradiction, which in terms of ethics implies honesty and truthfulness; and these qualities are valid for all civilized nations and for all time; so that this scientific principle may claim to rank among the first and most important of virtues. I do not think that I exaggerate in saying that an infraction of this ethical demand is discovered and repudiated more quickly and certainly in physics than in any other science.

It is rather shocking to notice the difference between such strictness and the thoughtless laxity with which similar faults are accepted in everyday life. I have not so much in mind the so-called conventional falsehoods which in practice are harmless and to a certain extent indispensable to daily intercourse: conventional falsehoods do not deceive precisely because they are conventional. The harm begins where there is an intention to deceive the other party and to convey to him a faulty impression. It is the duty of those who work in responsible positions to reform this matter ruthlessly as well as to set an example worth following.

Justice is inseparable from truthfulness: justice, after all, simply means the consistent application in practice of the ethical judgments which we pass on opinions and actions. The laws of nature remain fixed and unchanged whether applied to great or to small phenomena, and similarly the communal life of men requires equal right for all, for great and small, for rich and poor. All is not well with the State if doubts arise about the certainty of the law, if rank and family are respected in the courts, if defenseless persons feel that they are no longer protected from the rapacity of powerful neighbors, and if the law is openly wrenched on grounds of so-called expediency. The populace has a keen sense of the security of the law, and nothing rendered Frederick the Great more popular than the legend of the miller of Sans Souci. Such principles made Germany and Prussia great; it is to be hoped that they will never be lost, and it is the duty of every patriot to work for their preservation and consolidation.

At the same time it must be understood that the goal at which we aim – a permanently satisfactory condition – can never be attained in its perfection. The best and maturest ethical principles must fail to take us to an ideal perfection: they can never do more than indicate the direction in which we can look for our ideal. If these facts are disregarded there is a danger that the seeker may despair altogether or may doubt the value of ethics, a state in which, especially if he is honest in his dealings with himself, he may easily end by attacking ethics. There are numerous examples of this among the philosophies of ethics. The case here is the same as in science: what is important is not to have a permanent possession but to work unceasingly towards the ideal aim, to struggle daily and hourly towards a renewal of life, and despite every set-back to strive towards improvement and perfection.

Yet in the end we may be tempted to ask whether such an unceasing though fundamentally hopeless struggle is not wholly unsatisfactory. It may be asked whether a philosophy has any value at all if its votaries are left without a single fixed point affording them a firm and immediate security in the continual perplexity and hurry of their existence.

Fortunately this question admits of an answer in the affirmative. There is a fixed point and a secure possession which even the least of us can call his own at all times; an inalienable treasure which guarantees to thinking and feeling men their highest happiness, since it assures their peace of mind, and thus has an eternal value. This possession is a pure mind and good will. These afford secure holding ground in the storms of life and they are the primary condition underlying any really satisfactory conduct, as equally they are the best safeguard against the tortures of remorse. They are the essential of every genuine science and they are equally a sure standard by which to measure the ethical value of every individual.

> Those who are ever striving forward
> Them we can save.

2 Causality in Nature

Recent developments in physics have shown that the hopes of a more profound knowledge of nature which had been aroused by the brilliant successes of physical studies would have to be subdued in certain important points. It was seen, for example, that the law of causality could not possibly be applied universally in the customary classical form, since its application to the world of atoms had proved a definite failure. In consequence all persons interested in the meaning and significance of scientific study are compelled to examine afresh the essential quality of the laws of nature and, more especially, to scrutinize the concept of causality.

It is no longer possible to proceed as Kant did, who treated the law of causality as expressing the validity of invariable rules applicable to all events, and thus counted it among the categories, regarding it as a form of intuition without which experience would be impossible. No doubt the Kantian principle that certain categories are the *a priori* principles of all experience will remain unshaken for all times; yet this tells us nothing about the meaning of the individual categories; and the fact that the axioms of Euclidian geometry, which Kant treated as categories, have latterly proved not only capable, but actually in need of modification, has rendered physicists very cautious in this respect. In order, therefore, to proceed without prejudice, we must eschew dangerous assumptions and must begin by looking for a really reliable starting-point permitting us to introduce the concept of causality.

When we say that there is a causal connection between two consecutive events, we mean that there is some kind of law connecting them, the earlier event being called the cause, and the later the effect. The question then arises as to what is the specific nature of the nexus between them. Is there any criterion permitting us to say that a given natural event is the effect of another?

This question is as old as natural science itself, and the fact that it is continually being raised demonstrates that no definite answer has yet been found. This is unsatisfactory: but it becomes less so if we consider that it could not be otherwise. The hope that it might ever be possible first to produce an exact definition of causality, and then to use this definition as a basis for the investigation of the validity of the law of causality in nature, could only have been described as naive at an earlier time: today, in view of the developments which have taken place in the exact study of nature, it could only be described as foolish. In natural science, as in every other science, it is not the case that we begin from fixed fundamental concepts and then try to find out whether they are realized in the surrounding world. The opposite is true. Without previous preparation or information we are placed at birth in the very middle of life, and in order to find our way through this life which is ours whether we want it or not, we try to introduce order into our experience. To do this we use the mental faculties given to us at birth in order to form certain concepts which may be applied to the events which we have experienced and are likely to experience in future. Clearly such a procedure implies arbitrariness and obscurity: innumerable facts in every branch of science bear witness to this. At this point it must suffice to point out that even in mathematics, the most exact of the sciences, the controversy about the origin and meaning of the fundamental concepts is more violent today than ever before. If such is the case with mathematics it can hardly be expected that it will be easy to define the concept of causality as applied to nature in a way that will commend itself to all times and all civilizations.

Yet thinking men have never ceased to show interest in the question of the nature and validity of the law of causality; this interest is rapidly growing at the moment; and the conclusion to which we are led is that causality is something fundamental. We suspect that it is ultimately independent of our senses and of our intelligence and is deeply rooted in that world of reality where a direct scientific scrutiny becomes impossible. For surely it will be admitted that even if the earth with all its inhabitants were to perish, the cosmic events would still continue to obey their causal laws, even though no human being were alive to test the meaning and justification of such a claim.

In any case there is only one method of apprehending the real nature of causality. This method is to begin with the world of data which we possess, i.e., our experiences, to generalize, to eliminate as far as possible all anthropomorphic elements and thus cautiously to elaborate an objective concept of causality.

The many attempts which have been made in this direction show us that the best approach to the concept of causality consists in attaching it to the capacity of foretelling future events which we have acquired and tested in daily experience. And indeed there is no better means of demonstrating the causal connection between two events than to show that the occurrence of the one event can regularly permit us to forecast the occurrence of the other. This much was known to the farmer in the story who made such a striking demonstration before the skeptics of the causal connection between artificial manure and the fertility of the soil. The skeptics refused to believe that the heavy yield of clover on the farmer's field was caused by artificial manure and tried to discover some other reason. Thereupon the farmer plowed in lines having the shape of letters and had them manured while leaving the rest of his field without manure. When the clover came up in the following spring all could plainly read in letters of clover: "This portion has been manured with gypsum."

I propose to commence the next stage with the simple and general proposition that an event is causally conditioned if it can be foretold with certainty. Of course I mean no more by this than that the possibility of correctly foretelling the future is a safe criterion of the presence of a causal connection; I do not mean that the two are identical. To take a familiar instance: during the day we can foretell the coming of night with certainty and we may hence infer that night has a cause; but we do not for this reason treat day as being the cause of night. On the other hand it frequently happens that we assume the existence of a causal nexus where it is wholly impossible to make a correct forecast. This applies, for example, to the weather. The unreliability of weather prophets has become proverbial: yet presumably there is no trained meteorologist who could not believe the atmospheric events to be causally determined. Thus the proposition with which we started is seen to possess no more than a provisional value: we must go considerably deeper in order to understand the real nature of the concept of causality.

With regard to weather forecasting the obvious reflection is that it is unreliable only because the object in question, viz., the atmosphere, is so extensive and complicated. If we take a small part of it, e.g., a liter of air, we are in a much better position to foretell correctly its behavior when reacting to such external influences as compression, heat, moisture, etc. We are acquainted with certain physical laws which enable us to foretell more or less exactly the results of measurements we may make in order to discover the effects of an increased pressure, a higher temperature, condensation, etc.

On further scrutiny, however, we reach a very remarkable

discovery. However simple the conditions which we select and however delicate our instruments, we shall never be able to calculate in advance the result of the measurement with absolute accuracy, i.e., so as to agree to all places of decimals with the number measured. There always remains an element of inaccuracy. This is not the case in purely mathematical calculations, e.g., when the square root of 2 is calculated, which can be stated with complete accuracy to any number of places. And what applies to mechanics and heat is true of all the branches of physics, e.g., of electrical and optical events.

The available facts accordingly compel us to admit that the state of affairs may be correctly summed up by saying that in no single instance is it possible accurately to predict a physical event.

If we place this fact in juxtaposition with the proposition from which we started previously, when it was said that an event is causally determined if it can be accurately predicted, we find ourselves faced with an inconvenient but inescapable dilemma. If we rigidly maintain our original proposition then nature does not present us with a single instance where it is possible to assert that there is a causal connection; if we insist that somehow room must be found for a strict causality then we are compelled in some respect to modify the proposition from which we started.

There are at present a number of physicists and philosophers who prefer the first alternative. These I propose to call the Indeterminists. They maintain that there is no genuine causality or law in nature, and that the illusion of their existence is due to the fact that certain rules are found to occur which are very nearly but not absolutely valid. In principle the indeterminist looks for a statistical foundation in every physical law, even in that of gravitation; all these laws are for him laws of probability, referring to averages drawn from numerous similar observations, claiming no more than an approximate validity for single observations and always admitting exceptions.

A good example of one of these statistical laws can be found in the manner in which the pressure exerted by a gas on the wall of the containing vessel depends on the density and temperature of the gas. The pressure exerted by the gas is caused by the continual impact of extremely numerous molecules flying about at high velocity irregularly and in all directions. If the total energy exerted by these impacts is calculated, it is found as a result that the pressure exerted on the wall of the containing vessel is very nearly proportional to the density of the gas and to the square of the average velocity of the molecules. Further, this calculation agrees to a satisfactory degree, with actual measurements, provided that temperature is regarded as a measure of molecular velocity.

The theory is directly confirmed if we study the temporary variations in pressure which are observed if we concentrate upon the pressure exerted on any very small portion of the wall of the container. If we consider such a portion – e.g., the billionth part of a square millimeter – it may occur that we find a considerable time elapsing before a molecule happens to hit this particular surface, while on the contrary two or even three may strike it in quick succession. It is all a matter of chance. In these circumstances it is, of course, impossible to claim that there is a constant pressure exerted by the gas: the pressure, on the contrary, is subject to irregular variations. The simple law of pressure is valid only for relatively extensive surfaces on which a very great number of molecules exert an impact; for here the irregularities cancel each other.

Variations of this kind caused by the irregular impact of molecules are observed everywhere where molecules in rapid motion are in contact with bodies easily set in motion. They can for example also be observed in the movements first described by Brown and called after him. These are the trembling movements executed by fine particles of dust suspended in a liquid and subject to the impacts of the molecules of the liquid. The fact that a very sensitive balance never attains rest but continually oscillates irregularly around the point of equilibrium, is another instance of this movement.

Various radioactive phenomena afford another example of statistical laws. A radioactive substance continuously emits a number of particles having a positive or a negative charge, a process due to the spontaneous decomposition of its atoms. When dealing with comparatively lengthy periods of time, we can fairly say that the emission is steady. When dealing with briefer periods, however, i.e., with those which do not much exceed the average interval between two consecutive emissions, we find that the process is entirely irregular.

Now the indeterminists deal with every physical law in the same way as that in which they deal with the laws of the gases and of radioactivity: they treat them as being in the last analysis a matter of contingency. For them nature is entirely a matter of statistics and it is their aim to build up physics on a calculus of probability.

In fact, however, physics has hitherto developed on the opposite assumption, and physicists have chosen the second of the two above-mentioned alternatives. In other words, in order to be preserved intact, the principle of causality, according to which an event is causally determined only if it can be accurately foretold, has been slightly modified. What has been done is to change the sense in which the term "event" is employed. Theoretical physics does not consider an individual measurement as an event, because such a measurement always contains accidental and unessential elements. By an event, physics means a certain merely intellectual process. It substitutes a new world in place of that given to us by the senses or by the measuring instruments which are used in order to aid the senses. This other world is the so-called physical world image; it is merely an intellectual structure. To a certain extent it is arbitrary. It is a kind of model or idealization created in order to avoid the inaccuracy inherent in every measurement and to facilitate exact definition.

It follows that every measurable magnitude, every length, every period of time, every mass, and every charge, has a two-fold meaning. It may be considered as the immediate result of the measurement, or it may be treated as applied to the model to which we give the name of physical world image. In the former case it can never be defined exactly, and consequently can never be represented by an exact figure; in the second case it can be denoted by definite mathematical symbols with which we can operate in accordance with exact rules. If we speak in physics of the height of a tower and use a trigonometrical equation for its calculation, we have in mind a perfectly defined magnitude; an actual measurement of the height, on the other hand, does not give us an exact magnitude. Thus the ideal height (which can always be calculated with perfect accuracy) is always something different from the actually measured height, and the same applies to the period of oscillation of a pendulum or to the brightness of an electric globe. Further, any universal constant, e.g., the velocity of light in space, or the charge of an electron, is not the same in the physical world image and in any actual measurement: in the former it is perfectly exact; in the latter it is not accurately defined. A clear and consistent distinction between the magnitudes of the world of the senses and the similarly designated magnitudes of the world image is indispensable if we wish to have a firm grasp of the matter. Without it any debate on this question will always lead to misunderstandings.

It is not therefore the case, as is sometimes stated, that the physical world image can or should contain only directly observable magnitudes. The contrary is the fact. The world image contains no observable magnitudes at all; all that it contains is symbols. More than this: It invariably contains certain components having no immediate meaning as applied to the world of the senses nor indeed any meaning at all, e.g., ether waves, partial oscillations, reference coordinates, etc. Such component parts may seem to be an unnecessary burden; yet they are adopted because the introduction of the world image brings with it one decisive advantage. This advantage consists in the fact that it permits a strict determinism to be carried through.

It is true that the world image fulfills no more than an auxiliary function. In the last analysis it is the events of the world of the senses that matter, and the desideratum is to calculate them in advance as exactly as possible. According to the classical theory the procedure is as follows. The object, e.g., a system of material bodies, is taken from the world of the senses and is symbolized in any measured state; in other words, it is transferred into the world image. As a result we obtain a physical structure in a certain definite initial state. The external influences acting upon the object in subsequent time are similarly symbolized in terms of the world image. As a result of this second step we obtained the external forces acting upon the structure; in other words, the liminal conditions. These data causally determine the behavior of the system for all time, and it can be calculated with absolute accuracy from the differential equations furnished by theory. In this way the coordinates and the velocities of all material points of the system are found to be perfectly definite functions of time. If now at any later point we translate the symbols used for the world image back into the world of the senses, the result we obtain is that a later event of the sense-world has now been connected with an earlier event of the sense-world, so that the latter can be used in order to allow us to make an approximate forecast of the former.

We can sum up then by saying that, while the forecast of any event in the sense-world is always subject to a certain inaccuracy, all the events of the physical world image happened in accordance with certain definite laws which we can formulate so that they are causally determined. Hence, the introduction of the physical world image enables us to substitute the inaccuracies inherent in the translation of the event from the world of the senses to the world image and back from the latter to the former for the inaccuracy inherent in forecasting an event of the sense-world. It is in this that the importance of the physical world image consists.

Classical theory has tended to disregard the inaccuracies due to this transference. It has concentrated upon applying causality to the events in the world image, and by this method has obtained its striking successes. It has even succeeded in discovering a satisfactory explanation compatible with a strict causality for the above-mentioned irregular variations in the pressure of a gas, or in the movements of molecules (Brown's movement). As for the indeterminists, these phenomena do not constitute a problem for them: they look for irregularity behind every rule and statistical laws afford them immediate satisfaction. Accordingly, they confine themselves to assuming that the collision between two molecules or the impact of a single molecule on the container is governed by statistical laws. Yet there is not really any valid reason for this assumption any more than the fact that the electrons gather on the surface of a conductor allows us to infer that the charge of any individual electron is at its surface. The determinists, on the other hand, look for a rule behind every irregularity, and it is their task to formulate a theory of the laws of the gases on the assumption that the collision between any two molecules is causally determined. The solution of this problem was the life-work of the great physicist, Ludwig Boltzmann, and it is one of the finest triumphs of theoretical investigation. It does not only lead to the proposition that the average energy of the oscillations about the point of equilibrium varies as the absolute temperature – a proposition confirmed by measurements – but it also permits us to calculate with remarkable accuracy the absolute number and mass of molecules impinging, e.g., upon a very sensitive balance, simply by measuring its oscillations.

This success and others of a similar kind seemed to warrant the hope that the world image of classical physics might on the whole fulfill the task assigned to it and that the inaccuracies remaining after the process of translation out of and back into the world of the senses would ultimately be rendered progressively insignificant as methods of measurement became increasingly accurate. This hope has been destroyed for good with the entry on the scene of Planck's quantum.

The quantum theory evolved originally from the radiation of light and heat; accordingly we may well begin at this point by dealing with the processes of radiation. Numerous facts allow us to regard it as proved that the energy in a beam of light of any given color does not move in a steady continuous stream, but progresses in individual parts called photons, the size of which depends exclusively upon the color of the light; these photons fly from their source in all directions with the velocity of light and to this extent behave in accordance with Newton's emanation theory. Where the light is intense the photons follow each other so densely that they are practically equivalent to a steady continuous stream; however, as the distance from their source increases the density of the ray decreases and the photons are less close to each other, like a jet of water which grows progressively thinner until it turns into a number of individual drops of a certain magnitude. The characteristic fact is that the photons (the "drops" of energy) do not grow smaller as the energy of the ray grows less; what happens is that their magnitude remains unchanged and that they follow each other at greater intervals.

Now it is easy to see that the application of causality to these events leads us to serious difficulties.

Let us take, for example, a ray of a given color falling upon a highly polished level sheet of glass. Part of the light will then be reflected and another part, say three times as much, will pass through the sheet. The ratio between these two parts does not depend upon the intensity of the light, or, in other words, upon the number of photons impinging on the glass. This much is shown by experience. Now if the number of impinging photons is large, e.g., a million, it is easy to state how many will be reflected and how many will penetrate: a quarter of a million will be reflected, and three-quarters of a million will penetrate. If, however, the ray of light is extremely weak, a single photon may impinge on the sheet, and then the question whether it will be reflected or will penetrate is, to say the least of it, a source of serious embarrassment. The easiest solution would be to divide it into four: but this is impossible.

But worse is to come. In the previous example we might find a way out by assuming that, while there was a temporary state of uncertainty, there might still be some hitherto unknown factor decisively influencing the photon in one sense or the other. The following case, however, seems to be entirely hopeless. It is a fact that certain colors are reflected by preference while others are allowed to penetrate by preference. When a white ray falls on the sheet the sheet looks colored in the reflected light and also in the penetrating light. The classical wave theory of light gives an entirely satisfactory explanation of this phenomenon by saying that the light reflected at the front of the sheet interferes with that reflected at the back, so that the two reflected rays strengthen or weaken each other in accordance as the wave crest of one ray coincides with the crest or the trough of the other ray. Now the wave lengths of different colors are different, so that there are differences for the different colors, and the differences thus calculated agree exactly with actual measurements. This phenomenon, too, can be observed with light of the least intensity.

What happens now when a single photon impinges on the sheet? The photon must interfere with itself, since otherwise its wave length could not exert any influence. For this purpose, however, it would have to separate into parts; and this is impossible. We see thus that this view is altogether untenable.

Mechanics is in the same position as optics, as far as the quantum theory is concerned. The smallest mass points, the electrons, are in the same condition as the photons: they interfere with each other. An electron having a given velocity in this respect resembles a photon of a given power; if it impinges upon a sheet of crystal at a certain angle it is either reflected by preference or passes through by preference according to its velocity, and a complete explanation of this phenomenon in all its details is afforded by considering the wave length corresponding to its energy. The path taken by the electron when impinging upon the sheet has therefore never been calculated, and indeed it cannot be calculated.

The fundamental difficulty of determining the place of an electron moving at a certain velocity is expressed in a general manner by the uncertainty relation originally formulated by Werner Heisenberg. This relation is characteristic of quantum physics and states among other things that the measurement of an electron's velocity is inaccurate in proportion as the measurement of its position in space is accurate, and vice versa. It is not hard to discover the reason. We can determine the position of a moving electron only if we can see it and in order to see it we must illuminate it, i.e., we must allow light to fall on it. The rays falling on it impinge upon the electron and thus alter its velocity in a way which it is impossible to calculate. The more accurately we desire to determine the position of the electron, the shorter must be the light waves employed to illuminate it, the stronger will be the impact, and the greater the inaccuracy with which the velocity is determined.

This much having been discovered it is clearly impossible even in principle to transfer into the world of the senses with any desired degree of accuracy the simultaneous values of the coordinates and of the velocities of material points such as we find them at the core of the world image of classical physics. This impossibility makes it difficult to apply a strict causality and has led certain indeterminists to claim that the law of causality as applied to physics has been definitely refuted. On closer consideration, however, it is seen that this conclusion rests upon a confusion between the world image and the world of sense; it is at any rate premature. It is far more natural to avoid the difficulty by another method, a method which has often rendered good services in similar cases and which consists in assuming that it is meaningless, with respect to physics, to ask for the simultaneous values of the coordinates and of the velocities of a material point or for the path of a photon of a given color. Evidently the law of causality cannot be blamed because it is impossible to answer a meaningless question; the blame rests with the assumptions which lead to the asking of the question, i.e., in the present case with the assumed structure of the physical world image. The classical world image has failed us and something else must be put in its place.

This has actually been done. The new world image of quantum physics is due to the desire to carry through a rigid determinism in which there is room for Planck's quantum. For this purpose the material point which had hitherto been a fundamental part of the world image had to lose this supremacy. It has been analyzed into a system of material waves, and these material waves are the elements of the new world image.

The world image of quantum physics stands in approximately the same relation to classical physics as Huygens' wave optics stand to Newton's corpuscular or ray optics. The latter meets a great many instances, but it fails in others; and similarly classical or corpuscular mechanics is now seen to be no more than a special instance of the more general wave mechanics. In place of the material point of the classical system an infinitely narrow parcel of waves is found, i.e., a system of numerous waves interfering with each other in such a way as to cancel each other everywhere in space except at the place occupied by the material point.

The laws of wave mechanics differ, of course, fundamentally from those of classical mechanics with its material points. It is an essential fact, however, that the magnitude which is characteristic for the material waves is the wave function, by means of which the initial conditions and the final conditions are completely determined for all times and places. Definite rules of calculation are available for this purpose; it is possible to employ Schroedinger's operators, Heisenberg's matrices, or Dirac's Q-numbers.

Thus the introduction of wave functions solves the difficulty mentioned above, which arose when we asked how a single electron behaved when impinging on a crystal. The question then was whether it was reflected or penetrated the sheet. The impinging electron cannot divide into several parts; the waves, however, which are substituted for it can do so, so that interference becomes possible between the waves reflected at the front and those reflected at the back. Hitherto such a process was entirely incomprehensible: now it occurs in accordance with laws which can be exactly formulated.

We see then that there is fully as rigid a determinism in the world image of quantum physics as in that of classical physics. The only difference is that different symbols are employed and that different rules of operating obtain. Accordingly the same happens in quantum physics as we saw previously happening in classical physics. The uncertainty in forecasting events in the world of the senses disappears and in its place we have an uncertainty with regard to the connection between the world image and the world of the senses. In other words, we have the inaccuracy arising from a transfer of the symbols of the world image to the sense-world and vice versa. The fact that physicists have been willing to put up with this double inaccuracy is an impressive demonstration of the importance of maintaining the rule of determinism within the world image. At the same time a critical observer may well consider the price paid for the preservation of strict causality to be rather high. A superficial consideration shows how wide is the distance between the world image and the sense-world of quantum physics, and how much more difficult it is in quantum physics to translate an event from the world image into the sense-world and vice versa. Things are no longer as simple as they were in classical physics. There the meaning of each symbol was entirely clear; the position, the velocity, and the energy of a material point could be established more or less directly by measurement, and there was no apparent reason why it should not be assumed that any remaining inaccuracy would eventually be reduced below any given limit in the course of the progressively growing accuracy of the methods of measurement. The wave function of quantum mechanics, on the other hand, affords us in the first instance no help at all for an interpretation of the world of the senses; and while the term wave is expressive and suitable, it must not be allowed to disguise the fact that its meaning in quantum physics is totally different from that which it formerly had in classical physics. In classical physics a wave is a definite physical process, a movement perceptible by the senses or an alternating electrical field admitting of direct measurements, whereas in quantum physics it really denotes no more than the probability that a certain state exists. When a photon or electron impinges on the sheet of crystal it is not these entities which are divided, and thus lead to the phenomena of interference; all that we have is the probability that the indivisible photon or electron is present. It is only when a vast number of photons or electrons are impinging that this magnitude denotes a perfectly definite number of photons or electrons.

Such considerations have caused the indeterminists to renew their attacks on the law of causality. In the present instance they have some reason for expecting a certain positive success, since all measurements must have a merely statistical significance so far as they relate to wave functions. Yet here again the champions of strict causality have the same means of escape as before. Once again they can assume that there is no definite meaning in inquiring after the significance of any given symbol of the world image of quantum physics (e.g., a material wave) unless it is stated at the same time how this significance is to be determined and what is the condition of the special measuring instrument used in order to apply the symbol to the world of the senses. It is customary for this reason to speak of the causal work of the measuring instrument employed, by which it is meant that the inaccuracy is due at any rate in part to the fact that the magnitude to he measured is connected by some kind of law with the means by which it is measured.

As a matter of fact every measurement, whatever the method of its employment, invariably interferes more or less with the event to be measured, as was seen above when we dealt with the electron in motion whose path is interfered with when it is illuminated, the interference varying with the intensity of the illumination, and the illumination being essential for the measurement. Accordingly, when a given material wave at various times corresponds to various events in the world of the senses, the reason is that the sensuous meaning of the material wave does not depend solely upon the wave itself but also depends on the reciprocal interference between the wave and the measuring instrument.

The above assumption gives a new development to the entire question, the further course of which is as yet uncertain. For now the indeterminists can fairly ask whether the concept of the causal influence exerted by the measuring instrument upon the measured event has any rational meaning at all, in view of the fact that we are acquainted with the event only by measuring it, so that every measurement brings about a fresh causal interference – in other words, a fresh disturbance of the event. Thus it looks as though it must be impossible to distinguish between the "event in itself" and the apparatus by which it is measured.

This objection does not, however, meet the case. Every experimental physicist is aware that there are indirect as well as direct methods and that in many instances where the latter failed, the former have rendered useful services. And it is even more important that a word should be said to refute a widespread and plausible opinion which holds that a problem in physics deserves to be examined only when it is certain in advance that it admits of a definite answer. If this rule had always been followed, the famous experiment made by Michelson and Morley in order to measure the so-called absolute velocity of the earth would never have been undertaken, and we might well be without the theory of relativity today. The problem of the earth's absolute velocity has for some time been seen to be somewhat insignificant: yet the trouble spent upon it has proved extremely useful for physics. It is all the more likely that it may prove worth while to pursue the problem of a strict causality, since this question is far from being settled and might prove more fruitful than any other question in physics.

The question then remains how we are to reach a decision. Clearly all that we can do is to adopt one of the two opposite views and to see whether it leads to useless or to fruitful results. To this extent it is satisfactory to see that the physicists who interest themselves in this problem tend to fall into two schools, one of which tends towards determinism while the other tends towards indeterminism. It would seem that at present the latter constitute the majority, although it is not easy to be certain and changes may well occur in course of time. There might also be room for a third party which might take up a kind of mediating position, treating certain concepts like those of electrical attraction, or gravitation, as possessing an immediate significance and as being subject to strict laws while assuming others, like those of the light wave or material wave, to have a merely statistical meaning for the world of the senses. Yet such a view might be considered unsatisfactory because of its lack of unity, so that for the moment I propose to leave it aside and to deal with the two completely consistent points of view.

When the indeterminist finds that the wave functions of quantum physics are simply statistical magnitudes his zeal is satisfied and he feels no impulse to ask further questions. Again, when dealing with radioactive processes, he is satisfied to find, e.g., that a given number of atoms of any radium combination decompose on an average per second, and he does not ask why one atom happens to be decomposing now while its neighbor may survive a thousand years. On the other hand a definite natural law like Coulomb's law of electrical attraction is an unsolved problem for him since he cannot rest satisfied with Coulomb's method of expressing the potential and is compelled to look for exceptions. He rests satisfied only when he has succeeded in establishing the degree of probability that the electrical force differs from Coulomb's value by a certain given amount.

The determinist's standpoint is diametrically opposite in each detail; he is satisfied with Coulomb's law of electrical attraction because it is entirely definite, but he recognizes the wave functions as magnitudes having a probable value only so long as the apparatus is disregarded by which the wave is produced or analyzed. Further, he looks for a strict law governing the relations between the properties of the wave functions and the events in the bodies standing in a relation of reciprocal causality with the wave. For this purpose he must, of course, study all these bodies as well as the wave function, and he must transfer not only the entire experimental apparatus used for the production of the material waves – high-tension battery, incandescent wire, and radioactive material – but also the measuring apparatus, the photographic plate, the ionization chamber and Geiger's counter together with all the events occurring therein into his physical world image: and he must treat all these objects as constituting one single field of study, as a complete totality.

Of course this does not constitute a settlement of the problem; the problem, on the contrary, has for the moment become all the more complicated. It is not permissible to cut the structure in pieces, nor is any external interference permitted under penalty of destroying its uniqueness, so that a direct study of it is altogether impossible. On the other hand, we are now in a position to make certain novel hypotheses with regard to the internal events and subsequently to check the consequences. The future will show whether any advance is possible on these lines, and at the moment we cannot clearly see in what direction the advance is likely to lead. It may, however, be regarded as certain that Planck's quantum constitutes an objective limit beyond which the physical measuring instruments we possess cannot reach, and which will prevent us for all time from understanding the full causality of the most delicate physical processes "in themselves," i.e., apart from their origin and their effects.

In a way it would seem that we have now reached the end of our consideration, in the course of which we found that a strictly causal way of looking at things – "causal" being taken in the modified sense explained above – is wholly compatible with modern physics although its necessity cannot be demonstrated either *a priori* or *a posteriori*. Yet even here an objection occurs calculated to prevent a convinced determinist from being entirely satisfied with the interpretation of causality here introduced. Indeed, the objection is more likely to appeal to a determinist than to other persons. Even though we should succeed in developing the concept of causality on the lines here described, it will permanently be vitiated by a grave and fundamental defect. We were enabled to carry through the determinist view of the universe only by substituting the physical world image for the immediate world of the senses. Now the world image is due to our imagination and is of a provisional and changeable character; it is an emergency concept, hardly worthy of a fundamental physical notion, and the question arises whether it might be possible to endow the concept of causality with a more deep and direct significance by making it independent of the introduction of an artificial human product. This could be done by applying it, not to the physical world image but immediately to the experiences of the world of the senses. We shall, of course, have to maintain our original proposition, to the effect that an event is causally determined if we can accurately predict it: otherwise we would be surrendering our principle, which was to begin solely from actual experience. At the same time we are also compelled to accept our second proposition, which was that it was never possible to predict any event. It follows, in the same way as we saw above, that the first proposition must be somewhat modified if we wish to retain causality in nature. So far everything remains unchanged. The possibility now, however, arises of substituting a different and in a sense a contrary modification for the one hitherto adopted.

What we modified above was the *object* of the prediction, i.e., the event. What we did there was to refer the events not to the immediately given world of the senses but to a fictitious world image, by which process we were enabled to achieve an exact determination of the events. Now it is equally possible to modify the *subject* of the prediction, i.e., the predicting intellect. Every prediction implies a predicting person. In the subsequent argument I propose to concentrate upon the predicting subject and to treat the immediately given events of the world of the senses as object. An artificial world image will not be introduced at all.

It is easy to appreciate that the accuracy of a prediction largely depends on the individuality of the predictor. To revert to a weather forecast: it makes all the difference whether tomorrow's weather is foretold by somebody who knows nothing about the atmospheric pressure, the direction of the wind, and the moisture and temperature of the air, or by a practical farmer who notes all these things and has a long experience beside, or finally by a trained meteorologist who has weather charts from every part of the world, with exact data apart from this local information. The forecasts made by this series of prophets will show a diminishing degree of inaccuracy. That being so, we are induced to assume that an ideal intellect having complete knowledge of today's physical events in all places should be in a position to foretell tomorrow's weather with complete accuracy. The same applies to every forecast of physical events.

Such an assumption implies an extrapolation, a generalization which can neither be proved nor disproved by logical processes, and which consequently can be judged, not in accordance with its truth, but only in accordance with its value. From this point of view the impossibility of foretelling an event with complete accuracy in any single instance, whether we assume the standpoint of classical or quantum physics, appears to be the natural consequence of the fact that man with his senses and his apparatus is himself a part of nature to whose laws he is subjected. An ideal intellect is not so bound.

It might be objected that this ideal intellect itself is only a product of our thoughts and that the thinking brain is composed of atoms obeying physical laws. This objection will not bear close investigation. It is certain that our thoughts can carry us beyond any natural law known to us and that we can imagine connections between events which go far beyond those obtaining in physics. If it is claimed that the ideal intellect can exist only in the human brain, and would vanish with the disappearance of the latter, then in order to be consistent it would also have to be claimed that the sun and the whole external world in general can only exist in our senses, since these are the only source of scientific cognition. Yet every reasonable person must be convinced that the sun's light would not be diminished in the least even if the whole of mankind were to perish.

For we must take care not to regard the ideal spirit as ranking with ourselves; we have no right to ask it how it acquires the knowledge enabling it to foretell exactly future events, since such inquisitiveness might well meet with the reply: "You resemble the spirit which you can grasp, you do not resemble me." If the inquirer should remain obstinate despite this answer and should insist that the notion of an ideal spirit if not illogical, is at any rate void of content and superfluous, then we may fairly reply that a proposition is not scientifically valueless merely because it lacks logical foundation, and that such narrow formalism obstructs the source from which men like Galileo, Kepler, Newton, and many other great physicists drew their scientific inspiration. Consciously or unconsciously a devotion to science was a matter of faith for these men; they had an unshakable faith in a rational order of the world.

At the same time such a belief is not compulsory: we cannot order men to see the truth or prohibit them from indulging in error. Yet the simple fact that we are enabled, if only to a limited extent, to subject future natural events to our intellectual operations, and to guide them in accordance with our will, would necessarily remain a wholly unintelligible mystery if it did not allow us to have, at any rate, a premonition of a certain harmony between the outer world and the human spirit. Logically the extent which we attribute to the realm of this harmony is a question of secondary importance. The most perfect harmony and consequently the strictest causality in any case, culminates in the assumption that there is an ideal spirit having a full knowledge of the action of the natural forces as well as of the events in the intellectual life of men; a knowledge extending to every detail and embracing present, past, and future.

It may be asked what becomes of human free will on this assumption, and it may be suspected that by it man is degraded to the rank of a mere automaton. The question is a natural one, and though I have had various opportunities of dealing with it, it is so important that I am unwilling to let the present opportunity pass without briefly dealing with it. In my opinion there is not the slightest contradiction between the domination of a strict causality in the sense here adopted and the freedom of human will. The fact is that the principle of causality on the one hand and free will on the other refer to totally different matters. We saw above that we must assume the existence of an ideal and omniscient spirit if a strict causality is to be upheld in physical events; on the other hand, the question of free will is one for the individual consciousness to answer: it can be determined only by the ego. The notion of human free will can mean only that the individual feels himself to be free, and whether he does so in fact can be known only to himself. Such a state of affairs is entirely compatible with the fact that his motives could be comprehended in every detail by an ideal spirit. A feeling that such a state of affairs is derogatory to the ethical dignity of the individual implies an obliviousness of the vast difference between the ideal spirit and the intelligence of the individual.

Perhaps the most impressive proof that the individual will is independent of the law of causality will be found if the attempt is made to determine in advance the subject's own motives and actions on the sole basis of the law of causality – by a method of intense introspection. Such an attempt is condemned to failure in advance because every application of the law of causality to the will of the individual and every information gained in this way is itself a motive acting upon the will, so that the result which is being looked for is continually being changed. Hence it would be a complete mistake to attribute the impossibility of forecasting the subject's actions on purely causal lines to a lack of knowledge which might be overcome if the individual intelligence were suitably increased. Such an inference is analogous to the process of ascribing the impossibility of simultaneously determining exactly the position and the velocity of an electron to the inadequacy of our methods of measuring. The impossibility of foretelling the subject's actions on purely causal lines is not based on any lack of knowledge, but on the simple fact that no method by whose application the object is essentially altered can be suitable for the study of this object.

In consequence intellectual man can never have recourse to the principle of causality to determine his acts of will; for this purpose he must refer to a totally different law, namely, the law of ethics, which is based on a different foundation and cannot be comprehended solely by scientific methods.

Scientific thought always requires a certain distance and a clear separation as between the thinking subject and the object of his thought, and this distance is best guaranteed by the assumption of an ideal spirit. Now such a spirit can only be subject and can never be object.

It may be said that it constitutes an unsatisfactory negation if we are prohibited from making the ideal spirit the object of our thoughts; and it may be added that this may be too high a price to pay for a rigorous determinism. Yet the price is not as dear as that which the indeterminists have to pay in order to carry through their view of the universe; for these thinkers are compelled to set a limit to their impulse for knowledge at a much earlier stage, since they renounce the attempt to set up laws valid for individual cases – a degree of resignation so surprising that one asks how it comes about that so many physicists have declared their allegiance to the doctrine of indeterminism. The explanation, unless I am mistaken, is of a psychological nature. On each occasion when a new idea of any importance is brought forth in science, it is tested in every direction, and if it is found valuable the attempt is made to make it the foundation of an intellectual system as comprehensive and as self-contained as possible. Such was the fate of the theory of relativity and such is the present condition of the quantum theory. At its present stage quantum physics has culminated in the doctrine of wave functions and for this reason there is a tendency to assign a certain definitive significance to the wave functions. Now the wave function in itself is no more than a probable magnitude, and accordingly attempts are made to represent the search for this probability as being an ultimate and supreme task. In this way the concept of probability is made the ultimate foundation of the whole of physics.

I think it unlikely that this manner of formulating the question will continue to satisfy in the future. Even in the intellectual sphere, where the laws enunciate probabilities to a much greater extent than do the laws of physics, no individual event is considered as fully and scientifically explained until light has been thrown on its causal origin; it is much less probable that it will prove possible to continue to eliminate the question of causality in the sphere of the natural sciences.

It is true that the law of causality cannot be demonstrated any more than it can be logically refuted: it is neither correct nor incorrect; it is a heuristic principle; it points the way, and in my opinion it is the most valuable pointer that we possess in order to find a path through the confusion of events, and in order to know in what direction scientific investigation must proceed so that it shall reach useful results. The law of lays hold of the awakening soul of the child and compels it continually to ask why; it accompanies the scientist through the whole course of his life and continually places new problems before him. Science does not mean an idle resting upon a body of certain knowledge; it means unresting endeavor and continually progressing development towards an aim which the poetic intuition may apprehend, but which the intellect can never fully grasp.

3 SCIENTIFIC IDEAS: THEIR ORIGIN AND EFFECTS

It will be well to begin with some words of explanation on the subject of the present chapter. The origin and effect of scientific ideas may seem a somewhat general and also a somewhat arrogant theme; it might even be suggested that it would have been better had I confined myself to the ideas of natural science. Yet if I had so confined myself the ideas with which I propose to deal would have been restricted in a manner which I consider unnecessary and unnatural. Looked at correctly science is a self-contained unity; it is divided into various branches, but this division has no natural foundation and is due simply to the limitations of the human mind which compel us to adopt a division of labor. Actually there is a continuous chain from physics and chemistry to biology and anthropology and thence to the social and intellectual sciences, a chain which cannot be broken at any point save capriciously. Again, the methods used in the various branches are found, if closely considered, to have a strong inner resemblance, and if they appear to differ, it is only because they have to be adapted to the different subjects which they treat. This inner resemblance has become more and more evident in recent times, to the great advantage of the whole of science. Hence I consider myself entitled to begin with considerations applying to the whole of science; although of course when I pass to more particular applications I shall tend to confine myself to my own subjects. Let me begin by asking how a scientific idea arises and what are its characteristics. In asking these questions I cannot attempt, of course, to analyze the delicate mental processes taking place in the investigator's mind and, what is more, largely in his subconscious mind. These processes are mysteries which can be revealed only to a limited extent if at all, and it would be equally foolish and rash to attempt any study of their inmost nature. The most that we can do is to begin with the obvious facts, which means that we investigate those ideas which have actually proved their leavening force for any branch of science; and this in turn means that we ask in what form they first occurred and what was their content at that time.

The first result of such an investigation is the discovery of the following rule: any scientific idea arising in the mind of a scholar is based on a concrete experience, a discovery, an observation, or a fact of any kind, whether it is a physical or an astronomical measurement, a chemical or a biological observation, a discovery among the archives or the excavation of some valuable relic of an earlier civilization. The content of the idea consists in this experience being compared and being brought into contact with certain different experiences in the mind of the scholar, in other words, in the fact that it establishes a link between the old and the new, so that a number of facts which had hitherto co-existed loosely are now definitely interrelated. The idea becomes fruitful and hence attains value for science if the interconnection thus established can be applied more generally to a series of cognate facts: for the establishment of an interconnection creates order, and order simplifies and perfects the scientific view of the universe. What is most important, however, is that the task of applying the new idea in its entirety shall lead to new questions and hence to new studies and to new successes. And this is true of the physicist's hypotheses no less than of the interpretations established by the philologist.

I propose now to exemplify the above in some detail, and in doing so I desire to confine myself to my own subject of physics. The angle of vision may appear somewhat restricted; on the other hand I shall be able to throw a clearer light upon the subject.

A classical example of the sudden emergence of a great scientific idea is found in the story of Sir Isaac Newton who, sitting under an apple tree, was reminded by a falling apple of the movement of the moon around the earth and thus connected the acceleration of the apple with that of the moon. The fact that these two accelerations are to each other as the square of the radius of the moon's orbit is to the square of the earth's radius, suggested to him the idea that the two accelerations might have a common cause and thus provided him with a foundation for his theory of gravitation.

Similarly, James Clerk Maxwell, on comparing the strength of a current measured electromagnetically with the strength of a current measured electrostatically, found that the ratio between these two magnitudes agreed numerically with the speed of light, and thus formed the idea that electromagnetic waves are of the same nature as light waves. This agreement became the starting-point of his electromagnetic theory of light.

We thus find that it is a characteristic of every new idea occurring in science that it combines in a certain original manner two distinct series of facts; and this can be traced in every instance, though certain differences occur with regard to content and formation. These differences in turn bring about differences in the effect and the fate of the different scientific ideas. Some of them eventually become the common property of science, are taken for granted and cease to be stressed. Such has been the fate of the two ideas just mentioned: of Newton's idea about the similarity between the acceleration of the moon and the gravitational acceleration on earth; and of Maxwell's idea about the electromagnetic nature of light. It is true that a good deal of time had to elapse before the latter idea won acceptance; at first, it tended to be disregarded, especially in Germany, where Wilhelm Weber's theory, which was based on the assumption of immediate action at a distance, held the stage. It was not until Heinrich Hertz made his brilliant experiment with ultra rapid electric oscillations that Maxwell's theory obtained the recognition it deserved.

Other ideas which have become the lasting heritage of science are those which hold that sound waves are of a mechanical nature and that rays of light and heat are identical. Teachers of physics tend to deal all too briefly with these ideas, and they should be reminded that there was a time when these ideas were far from being common-places. The second of the two just mentioned was indeed for years the subject of fierce controversy. It may be mentioned as a curiosity that the scientist whose experiments contributed most to its success – the Italian physicist, Macedonio Melloni – began by being one of its opponents, an instructive example showing that scientific values are independent of their theoretical interpretation.

But most of the ideas which play a part in science are different from those enumerated. The latter were perfect when they first took shape and will always retain their validity unchanged; these others assume their final form gradually, retain their value for a time and eventually either die or are modified to a more or less considerable degree. Frequently enough they resist modification and this resistance tends to be obstinate in proportion to their past successes: there have been occasions when this resistance has sensibly hampered the progress of science. Physics offers some instructive examples which it may be worth while to discuss in detail.

I propose to begin with the idea of the nature of heat.

The first stage in the development of the theory of heat consisted in calorimetry. It was based on the assumption that heat behaves like a delicate substance which flows from the hotter to the colder body whenever there is contact between two bodies having different temperatures. No quantitative change is supposed to take place during this process. This hypothesis worked well so long as no mechanical effects entered into play. A difficulty consisted in the production of heat by friction or compression, and this it was sought to overcome by assuming that the capacity of bodies for heat was variable, so that heat could be pressed out of a body under compression, like water being pressed out of a wet sponge, during which process the quantity of water remains unchanged. Later, when the invention of heat-utilizing power systems made more urgent the question of the laws governing the production of mechanical work from heat, Sadi Carnot tried to formulate the production of work out of heat on the analogy of the production of work out of gravity. As the falling of a weight from a greater to a less height can produce work, so the transition from a higher to a lower temperature can be used for the same purpose; and as the work obtained from gravitation varies as the weight of the body and the difference in height, so the work produced by heat varies as the amount of heat transferred and the difference in temperature. This materialist theory of heat received a shock from the empirical fact that a body's capacity for heat remains practically unaffected by compression and by friction; and it was finally refuted by the discovery of the mechanical heat equivalent, the significance of which consists in the fact that heat is dissipated in friction and new heat is produced in compression. The older theories of heat were thus reduced *ad absurdum* and it became necessary to build up a new theory. This task was undertaken by Rudolf Clausius and it was fulfilled in a number of classical works in which the second main principle of thermal dynamics was established. This principle presupposes that there are irreversible processes, i.e., processes which cannot in any way whatever be reversed. Now the conduction of heat, friction, and diffusion are among these processes.

Carnot's theory to the effect that the transition from a higher to a lower temperature was analogous to the falling of a weight from a higher to a lower level was not, however, to be so easily refuted. There were physicists who considered Clausius' ideas unnecessarily complicated and vague and who objected particularly to the introduction of the idea of irreversibility, by which a unique position among the various kinds of energy was assigned to heat. Accordingly they formed the theory of energetics in opposition to Clausius' thermo-dynamics. The first principle of this theory agrees with that of Clausius in enunciating the preservation of energy; the second principle, however – that which indicates the sense of events – postulated a thorough-going analogy between the transition from a higher to a lower temperature and the falling of a weight from a higher to a lower level, or again, the passing of electricity from a higher to a lower potential. Hence it came about that irreversibility was declared superfluous in order to prove the second principle, and that the existence of an absolute zero was denied, it being pointed out that temperature resembled levels of height and levels of potential in that only differences and nothing absolute could be measured. The fundamental distinction which consists in the fact that a pendulum swings past the position of equilibrium before coming to rest and that a spark passing between two conductors having opposite charges oscillates, whereas there is no such thing as an oscillation of heat between two bodies between which heat is passing, was considered irrelevant by the energetist school and was passed over in silence.

I myself experienced during the 80's and 90's of the last century what the feelings of a student are who is convinced that he is in possession of an idea which is in fact superior, and who discovers that all the excellent arguments advanced by him are disregarded simply because his voice is not powerful enough to draw the attention of the scientific world. Men having the authority of Wilhelm Ostwald, Georg Helm, and Ernst Mach were simply above argument.

The change originated from a different side altogether: atomism began to make itself felt. The atomic idea is extremely old; but its first adequate formulation took shape in the kinetic gas theory which originated more or less contemporaneously with the discovery of the mechanical heat equivalent. The energetists at first opposed it vigorously, and it led a modest existence; towards the end of last century, however, experimental investigation led to its rapid success. According to the atomist idea the transference of heat from the hotter to the colder body does not resemble the falling of a weight; what it resembles is a mixing process, as when two different kinds of powder in a vessel, having first constituted different layers, eventually mingle with each other if the vessel is continually shaken. If this happens the powder does not oscillate between a state of complete mixture and complete isolation of the constituent powders; what happens is that the change takes place once in a certain sense, viz., in the direction towards complete mixture, and is then at an end: the process is an irreversible one. Seen in this light the second principle of thermo-dynamics is found to be of a statistical nature: it states a probability. The arguments supporting this view and indeed raising it beyond any doubt have been well stated by my colleague, Max von Laue.

The historical development here described may well serve to exemplify a fact which at first sight might appear somewhat strange. An important scientific innovation rarely makes its way by gradually winning over and converting its opponents: it rarely happens that Saul becomes Paul. What does happen is that its opponents gradually die out and that the growing generation is familiarized with the idea from the beginning: another instance of the fact that the future lies with youth. For this reason a suitable planning of school teaching is one of the most important conditions of progress in science. Accordingly, I should like to deal briefly here with this point.

What is learned at school is not as important as how it is learned. A single mathematical proposition which is really understood by a scholar is of greater value than ten formulae which he has learned by heart and even knows how to apply, without, however, having grasped their real meaning. The function of a school is not so much to teach a business-like routine as to inculcate logical and methodical thought. It may be objected that ultimately it is the ability to do things rather than knowledge that matters; and it is true that the latter is valueless without the former, just as any theory is ultimately important only by reason of its particular applications. Yet routine can never be a substitute for theory, for in any cases that fall outside the rule, routine breaks down. Hence the first requisite, if good work is to be done, is a thorough elementary training; and here it is not so much the quantity of facts learned as the manner of treatment that matters. Unless this preliminary training is acquired at school, it is hard to obtain it at a later stage: training colleges and universities have other tasks. For the rest, the last and highest aim of education is neither knowledge nor the ability to do things, but practical action. Now practical action must be preceded by the ability to act, and the latter in turn demands knowledge and understanding. The present age, which lives at such a rapid rate, and shows so much interest for every innovation having an immediate sensational effect, provides us with instances where scientific training tends to anticipate certain exciting results before they have properly ripened; for the public is favorably impressed if the curriculum of an intermediate school already contains modern problems of scientific investigation. Yet such a practice is exceedingly dangerous. The problems cannot possibly be dealt with thoroughly, and the consequence may easily be to induce a certain intellectual superficiality and empty pride in knowledge. I should consider it extremely dangerous if the intermediate schools were to deal with the theory of relativity or the quantum theory. Specially gifted scholars always require exceptional treatment; but the curriculum is not designed for such, and I would definitely condemn any attempt to take such a question as that of the universal validity of the principle of the preservation of energy – which, of course, today is seriously regarded as an open one in nuclear physics – and to treat it as debatable before pupils who cannot have properly grasped the meaning of the principle involved, much less its potential scope.

The results of such an up-to-the-minute method of teaching become all too plain when we consider the way in which the breakdown of the exact sciences is occasionally spoken of today. It is characteristic of the prevalent confusion that there are numbers of inventive minds busying themselves to-day upon devices which aim at the unlimited production of energy or the utilization of the fashionable mysterious earth rays. And it is even more surprising that credulous persons provide ample funds for such inventors, while really valuable and hopeful scientific investigations are hampered or actually stopped by lack of means. A thorough school training might here prove a useful remedy, and this would apply to the patrons no less than to the inventors.

After this educational digression I should like to deal briefly with another physical idea whose varying fate may prove even more instructive than the changes undergone by the theory of heat. What I have now in mind is the idea of the nature of light.

The study of the nature of light began with the measurements of the speed of light. The idea which led Newton to his emanation theory established a comparison between a ray of light and a jet of water; the velocity of light was compared with the velocity of particles of water flying in a straight line. This hypothesis, however, failed to give an account of the phenomenon of light interference, i.e., of the fact that two rays of light meeting at a point can in certain circumstances produce darkness at this point. Accordingly the emanation theory was given up and its place was taken by Huygens' theory of undulations, where the underlying idea is that light is propagated like a wave of water which spreads concentrically in all directions from its point of origin at a velocity which, of course, is not connected in any way with the velocity of the particles of water. This theory succeeded completely in accounting for the phenomena of interference: two waves on impinging on each other can cancel each other whenever the crest of one wave impinges on the trough of another. However, this theory, too, did not last longer than a century. The undulation theory failed to explain the effect at a great distance of a ray of light having a short wave length. The intensity of light decreases as the square of the distance, so that if light is radiated equally in all directions it is impossible to understand how a ray is capable of producing, even at a very great distance, a quantity of energy which is entirely independent of its intensity, and which is relatively very considerable in the case of short waves like those of Röntgen rays or Gamma rays. Such powerful effects combined with extremely feeble intensity become intelligible only if we imagine the energy of light to be concentrated upon distinct, unchangeable particles or quanta. In a sense, this is a return to Newton's hypothesis of light particles.

At present, then, the position is an exceedingly unsatisfactory one. We have two theories facing each other like two equally powerful rivals. Each possesses keen weapons, and each has a vulnerable spot. It is hard to foretell the ultimate issue, but it is probably correct to say that neither theory will prove completely victorious. It is more likely that in the end a higher standpoint will be reached, where we shall be able to survey clearly the claims and the deficiencies of each of the two hypotheses. Such a standpoint can probably be found only if we intensify our search for the source of all experience, which would mean in the present case that we would turn our attention to the measurement of optical phenomena. This in turn would imply that we would turn our investigation upon the actual measuring instruments, a step which, in principle, is of enormous importance since it may be described as the introduction of totality into physics. According to this principle the laws of an optical phenomenon can be completely understood only if the peculiarities of the process of measurement are studied as well as the physical events at the points where the light originates and spreads. The measuring instruments are not merely passive recipients simply registering the rays impinging upon them: they play an active part in the event of measuring and exert a causal influence upon its result. The physical system under consideration forms a totality subject to law only if the process of measuring is treated as forming part of it.

How progress is to be made by this road is a difficult question and of much importance for the future. In order to appreciate its significance I propose to extend the scope of my survey, to go beyond the special conditions of optics and to approach the problem from a more general point of view.

Is it at all possible to predict with confidence the mutations of any scientific idea? Is it possible to claim that there is so much as an approximate law governing the development of scientific ideas? Looking back on the historical development of events one is tempted to suspect such a law, on considering that many important ideas began by existing in the dark, uncomprehended by the many and at best dimly foreseen by a few students who were in advance of their age; but that once mankind had become ripe for them, they came to life suddenly and simultaneously in a number of different places. The principle of the preservation of energy can be traced back for centuries in a rudimentary form; but it was not until the middle of last century that the principle was given a scientifically practical foundation, more or less simultaneously, by four or six students between whom there was no connection whatever. We may probably assert that even if Julius Robert Mayer, James Prescott Joule, Ludwig August Colding, and Hermann von Helmholtz had not been living at that time, the principle of the preservation of energy would, nevertheless, have been discovered only a little later. I would even venture to assert much the same of the origin of the modern theory of relativity or the quantum theory, were I not reluctant to face the obvious rejoinder that such prophecies after the event are somewhat cheap. I consider the inevitable element of such a process to consist in the fact that with the spread of experimentation and the improvements in methods of measurement, theoretical investigation has been forced in a certain direction almost automatically.

Yet there could be no greater mistake than to assume that the laws governing the growth and effect of scientific ideas can ever be reduced to an exact formula valid for the future. Ultimately any new idea is the work of its author's imagination, and to this extent progress is tied to the irrational element at some point even in mathematics, the most exact of the sciences; for irrationality is a necessary component in the make-up of every intellect.

If we bear in mind that any given idea is due to a given experience, we shall find it natural that the present time, so rich in numbers of new events, has proved a fruitful soil for the production and promulgation of new ideas. If, further, we consider that whenever an idea is formulated a relation is established between two different events, we shall find, even by the formal rules of combinations, that the number of possible ideas exceeds by an order of magnitude the number of available events.

Another circumstance explaining the vast output of scientific ideas at the present day possibly consists in the fact that owing to the spread of unemployment there are many lively intellects which experience a desire for productive work, and welcome a preoccupation with general theoretical and philosophical problems as a cheap and satisfactory escape from the emptiness of their everyday existence. Valuable results, unfortunately, are rare exceptions. I do not exaggerate when I say that hardly a week passes in which I do not receive one or more papers of varying length from members of every profession – teachers, civil servants, writers, lawyers, doctors, engineers, architects – with a request for my opinion. A thorough examination of these would take up all and more than all of my spare time.

These communications can be divided into two classes. The first is entirely naive and their authors have never considered that a new scientific idea to be valuable must be based on certain facts, so that specialized knowledge is essential for their formulation. The authors of these contributions, on the other hand, imagine that they have a fine prophetic gift enabling them to guess the truth direct, never suspecting that every important discovery is preceded by a period of hard individual work. These people, on the other hand, imagine that a happy fate has allowed the desired fruit to drop into their lap in the way in which Newton, sitting under the apple tree, received the idea of universal gravitation. What is worse is that these visionaries float above the surface, never penetrating to the depths, and are too ignorant scientifically to be capable of seeing their error. The dangers which flow from them should not be underestimated. It is satisfactory to note that modern youth shows a growing interest in general questions and in the acquisition of a satisfactory view of life; but for this very reason it should never be forgotten that such a view is baseless and doomed to sudden destruction unless it has a firm foundation in reality. Anyone desirous of obtaining a scientific view of the world must first acquire a knowledge of the facts.

Today the individual student can no longer form a comprehensive view of every department of science and in most instances he must take his facts at second hand. It is all the more important that he should be master of one trade and have an independent judgment on his own subject. Personally, as a member of the philosophical faculty, I have always asked that candidates for a philosophical doctorate should give evidence of special knowledge in one given special science. Whether this department belonged to the natural sciences or to the intellectual sciences is not important: what is important, is that the candidate should have acquired by actual study an idea of scientific method.

It is generally easy to demonstrate the worthlessness of the type of papers just mentioned; but there is another class which requires much more serious attention because the authors are careful students turning out excellent work in their special field. The scale of scientific work being such as it is today, specialization continually becomes more intense and consequently the more serious student experiences a desire to look beyond the limits of his own subject and to apply the knowledge acquired to other departments of science. There is thus a tendency to link two distinct departments by one idea which seems convincing to the student, who in this way transfers the laws and methods with which he has grown familiar within his own sphere to an alien one whose problems he thus tries to solve. There is especially among mathematicians, physicists, and chemists, a tendency to employ their own exact methods in order to throw light on biological, psychological, and sociological questions. Yet it must not be forgotten that such a new intellectual bridge to be sound requires both its pillars to be securely founded: it cannot fulfill its purpose unless the further pillar, too, has a proper foundation. In other words, it does not suffice for an ingenious student to be thoroughly acquainted with his original subject; if his more widely ranging ideas are to be fruitful, he must also have some knowledge of the facts and problems of the other sphere to which he is applying his idea. This deserves all the more emphasis because every expert tends to exaggerate the importance of his special field in proportion to the length of time spent on it and to the difficulties encountered. And once he has discovered the solution of a problem, he tends to exaggerate its scope and to apply the solution to cases of a totally different nature. Those who feel the desire to take up a higher standpoint than that which their own restricted field allows them, should never forget that there are students at work in other departments of science who are working with equal care and under equal difficulties although with different methods. The history of every science shows how frequently this rule is disregarded. In selecting my examples, however, I shall take care to confine myself to physics in order to avoid the mistake I have just been criticizing.

Among the more general ideas of physics there is practically none which has not been transferred with more or less skill to some other sphere by means of some association of ideas, an association depending frequently enough merely upon such contingent externals as terminology. Thus the term "energy" leads students to apply the physical concept of energy and with it the physical proposition enunciating the preservation of energy to psychology, and serious attempts have been made to subject the cause and degree of human happiness to certain mathematically formulated laws. The same must be said of attempts to apply the principle of relativity outside physics, e.g., in esthetics, or even in ethics. Yet there could be nothing more misleading than the meaningless statement that everything is relative. The proposition does not apply even in physics. All the so-called universal constants – the mass or the charge of an electron or a proton, or Planck's quantum – are absolute magnitudes: they are the fixed and unchangeable components of which the structure of atomism is built up. Of course a magnitude which once was considered absolute has often been found to be relative later; but whenever this happened another and more fundamental absolute magnitude was substituted. Unless we assume the existence of absolute magnitudes no concept can be defined and no theory can be formed.

The second principle of thermo-dynamics, the principle of the increase of entropy, has frequently been applied outside physics. For example, attempts have been made to apply the principle that all physical events develop in one sense only to biological evolution, a singularly unhappy attempt so long as the term evolution is associated with the idea of progress, perfection, or improvement. The principle of entropy is such that it can only deal with probabilities and all that it really says is that a state, improbable in itself, is followed on an average by a more probable state. Biologically interpreted, this principle points towards degeneration rather than improvement: the chaotic, the ordinary, and the common is always more probable than the harmonious, the excellent, or the rare.

Besides the misleading ideas which we have been considering there is another class which consists of those ideas which, looked at carefully, are seen to have no meaning at all. These play a fairly important part in physics, too. A comparison between the movement of an electron around a proton and the movement of a planet around the sun has caused investigators to study the velocity of the electron, although later investigation showed that it is completely impossible to answer these two questions simultaneously. Once again we see the danger of applying ideas and propositions which have proved their value in one department of science to another, and we perceive how great is the need of care in testing and formulating a new idea.

Yet there is also a theoretical side to the matter, of which it is now high time to speak. If a new idea were to be admitted only when it had definitely proved its justification, or even if we merely demanded that it must have a clear and definite meaning at the outset, then such a demand might gravely hamper the progress of science. We must never forget that ideas devoid of a clear meaning frequently gave the strongest impulse to the further development of science. The idea of an elixir of life or of the transmutation of base metals gave rise to the science of chemistry; that of perpetual motion to an intelligent comprehension of energy; the idea of the absolute velocity of the earth gave rise to the theory of relativity, and the idea that the electronic movement resembled that of the planets was the origin of atomic physics. These are indisputable facts, and they give rise to thought, for they show clearly that in science as elsewhere fortune favors the brave. In order to meet with success it is well to aim beyond the goal which will eventually be reached.

Looked at in this light the ideas of science wear a new aspect. We find that the importance of a scientific idea depends, frequently enough, upon its value rather than on its truth. This applies, e.g., to the concept of the reality of an external world or to the idea of causality. With both the question is not whether they are true or false, but whether they are valuable or valueless. This fact will appear all the more striking if we consider that the values of an objective science like physics are, to start with, wholly independent of the objects to which they relate; and the question arises how it comes about that the importance of a physical idea can be fully exploited only if we take its value into consideration.

In my opinion the only possible method available here is that which we followed when dealing with optics, a method applicable not only to physics, but to every department of science. We must go back to the source of every science, and we do this when we remember that every science requires some person to build it up and to communicate it to others. And this means once again the introduction of the principle of totality.

In principle a physical event is inseparable from the measuring instrument or the organ of sense that perceives it; and similarly a science cannot be separated in principle from the investigators who pursue it. A physicist who studies experimentally some atomic process interferes with its course in proportion as he penetrates into its details, and the physiologist who subdivides a living organism into its smallest parts injures or actually kills it; by the same token the philosopher, who in examining a new idea confines himself to asking to what extent its meaning is evident *a priori*, hampers the further development of science. Hence a positivism which rejects every transcendental idea is as one-sided as a metaphysics which scorns individual experience. Each method has its justification, and each can be carried through consistently; but if carried to an extreme they paralyze the progress of science because they prohibit the asking of certain fundamental questions, although they do so for opposite reasons: positivism, because the questions are meaningless, and metaphysics, because the answer to them is already available. The rivalry between the two parties will never be decided in favor of either, and in the course of history success has always wavered between the two. A century ago metaphysics enjoyed a hegemony which was followed by a melancholy collapse. Today positivism is striving after the leading position, which it will fail to obtain just as metaphysics failed.

Nobody had a deeper sense of this persistent antagonism than Goethe, who struggled with it all his life and has given it masterly expression in a number of different forms. He tried to overcome this antagonism by rising to the concept of totality, the introduction of which does justice to both views. Yet even Goethe's all-embracing mind was subject to the limits of time; he declined to admit the distinction between the rays of light in external space and the sensation of light in consciousness, and hence was prevented from doing justice to the brilliant progress made by physical optics in his time. Nevertheless, on observing the modern introduction of the idea of totality in physics, he might see in this change a confirmation of his way of thought.

Thus we observe, what we have already observed on several occasions, that there is an irrational core at the center of science which no intelligence can solve, and which no modern attempt at limiting by definition the tasks of science can remove. At first such a state of affairs may appear strange and unsatisfactory; on reflection, however, it will be seen that it could not be otherwise. For a close examination will show that every science really tackles its task at the center and not at the beginning, and that it is compelled to grope its way more or less laboriously towards the beginning without any hope of ever quite reaching it. Science does not find ready-made the concepts with which it operates: it has to form them artificially and their perfecting is a gradual process. It draws its material from life and it reacts upon life; its impulse, its consistency, and its vitality came from the ideas at work in it. It is the ideas which place before the student the problems with which he deals, which impel him to work without cease, and which enable him correctly to interpret the results he obtains. Without ideas investigation becomes aimless and the energy expended upon it is wasted. Ideals alone make a physicist of an experimenter, an historian of a chronicler, and a philologist of a graphological expert. We have already seen that the truth or falsity of an idea and the question whether it has a definite meaning is relatively unimportant: what matters is that it shall give rise to useful work. In science, as in every other sphere of cultural development, it is the work done which is the sole certain criterion of the health and the success of the individual as well as of the community. Accordingly, I wish to conclude these observations on the growth and effect of scientific ideas by quoting words in praise of work as applied to science; words which the Association of German Engineers, justly appreciating its theoretical and practical value, has made into its motto: "What is needed is investigation."

4 SCIENCE AND FAITH

A vast volume of experiences reaches each one of us in the course of a year; such is the progress made in the various means of communication that new impressions from far and near rush upon us in a never-ending stream. It is true that many of them are forgotten as quickly as they arrive and that every trace of them is often effaced within a day; and it is as well that it should be so: if it were otherwise modern man would be fairly suffocated under the weight of different impressions. Yet every person who wishes to lead more than an ephemeral intellectual existence must be impelled by the very variety of these kaleidoscopic changes to seek for some element of permanence, for some lasting intellectual possession to afford him a *point d'appui* in the confusing claims of everyday life. In the younger generation this impulse manifests itself in a passionate desire for a comprehensive philosophy of the world; a desire which looks for satisfaction in groping attempts turning in every direction where peace and refreshment for a weary spirit is believed to reside.

It is the Church whose function it would be to meet such aspirations; but in these days its demands for an unquestioning belief serve rather to repel the doubters. The latter have recourse to more or less dubious substitutes, and hasten to throw themselves into the arms of one or other of the many prophets who appear preaching new gospels. It is surprising to find how many people even of the educated classes allow themselves to be fascinated by these new religions – beliefs which vary from the obscurest mysticism to the crudest superstition.

It would be easy to suggest that a philosophy of the world might be reached from a scientific basis; but such a suggestion is usually rejected by these seekers on the ground that the scientific view is bankrupt. There is an element of truth in this suggestion, and, indeed, it is entirely correct if the term science is taken in the traditional and still surviving sense where it implies a reliance on the understanding. Such a method, however, proves that those who adopt it have no sense of real science. The truth is very different. Anyone who has taken part in the building up of a branch of science is well aware from personal experience that every endeavor in this direction is guided by an unpretentious but essential principle. This principle is faith – a faith which looks ahead. It is said that science has no preconceived ideas: there is no saying that has been more thoroughly or more disastrously misunderstood. It is true that every branch of science must have an empirical foundation: but it is equally true that the essence of science does not consist in this raw material but in the manner in which it is used. The material always is incomplete: it consists of a number of parts which however numerous are discrete, and this is equally true of the tabulated figures of the natural sciences, and of the various documents of the intellectual sciences.

The material must therefore be completed, and this must be done by filling the gaps; and this in turn is done by means of associations of ideas. And associations of ideas are not the work of the understanding but the offspring of the investigator's imagination – an activity which may be described as faith, or, more cautiously, as a working hypothesis. The essential point is that its content in one way or another goes beyond the data of experience. The chaos of individual masses cannot be wrought into a cosmos without some harmonizing force and, similarly, the disjointed data of experience can never furnish a veritable science without the intelligent interference of a spirit actuated by faith.

The question now arises whether this deeper view of the various sciences can provide us with a philosophy of the world fit to be applied to the problems of life. The best answer to this question is furnished by reference to certain great scientists who accepted this view and who, in fact, found that it rendered them this service. Among many other investigators whose straitened existence was rendered supportable and even illustrious by science, I would mention in the first place Johann Kepler. Looked at from without the whole of his life was hampered by penury, disappointments, and distress: he was "by poverty oppressed": in the last year of his life he was compelled to appeal to the Diet at Regensburg for payment of the imperial pension, then long overdue. Perhaps his greatest trial came when he was forced to defend his mother against a charge of witchcraft. And what supported him in all this trouble, and rendered him capable of work, was the science he served: not the figures relating to his astronomical observations, but the belief which he drew from them in the rule of rational laws in the universe. It is instructive to compare his case with that of his master and chief, Tycho Brahe. The latter had the same scientific knowledge and disposed of the same observed facts: what he lacked was faith in the eternal laws and so it came about that Tycho Brahe remained one meritorious investigator among others, while Kepler became the founder of modern astronomy.

Another name occurring in this connection is that of Julius Robert Mayer, the discoverer of the mechanical heat equivalent. Mayer was not oppressed by financial troubles as Kepler was; but he suffered all the more from the neglect of his theory of the conservation of energy: in the middle of last century every scientist displayed the greatest suspicion of everything that had a flavor of natural philosophy. Yet Mayer remained undismayed by the silence with which he was met, and found consolation not so much in his knowledge as in his faith. In the end he lived to find the representatives of his department of science – the Society of German Naturalists and Physicists, among them Hermann Helmholtz – giving public expression to the recognition which had so long been denied him.[1] We find then in these and many similar instances an active faith at work, and we see that this faith is the power which gives their real effectiveness to the individual data of science. We may even go a step further and claim that a prophetic faith in the deeper harmony can render valuable services at the earliest stage – the stage of gathering the data. This faith points the way and sharpens the senses. An historian looking for documents in the archives and studying what he discovers, or an experimenter who pursues his work in the laboratory and scrutinizes his results, frequently finds the progress of his work facilitated – more especially when he comes to distinguish essentials from unessentials – if he possesses a more or less deliberate intellectual attitude which guides his investigations and serves to interpret the results. His experience then resembles that of a mathematician who discovers and formulates a new proposition before he can prove it.

There still remains a danger, and one which is perhaps the gravest that can lie in wait for the investigator. It should not be passed over in this connection. It consists in the fact that the given data may be falsely interpreted or even ignored. If this happens, science becomes a falsehood, an empty structure which collapses at the first shock. Innumerable scientists, both young and old, have succumbed to this danger in their enthusiasm for a scientific conviction. The danger is as grave today as ever it was; and the only remedy against it consists in respect for facts. The more fruitful a thinker's imagina-

---

[1]This was at the Annual Meeting of 1869, at Innsbruck.

tion is, the more careful he should be never to forget that the different facts invariably form the foundation without which science cannot exist; and the more carefully must he ask himself whether he is treating them with due respect.

It is only when we have planted our feet on the firm ground which can be won only with the help of the experience of real life, that we have a right to feel secure in surrendering to our belief in a philosophy of the world based upon a faith in the rational ordering of this world.

INDEX

Printed in Great Britain
by Amazon

54364465R00051